W9-AQB-332

Praise for *Superlative*

"In *Superlative*, Matthew LaPlante takes us on a whiplash-paced journey around the globe to visit the biggest, smallest, quickest, slowest, and smartest creatures out there. In a string of short colorful vignettes, LaPlante explores a medley of superlative creatures one at a time, intertwining science and natural history with spirited storytelling and genuine affection. In the end, we learn that what makes each of these creatures superlative also makes them uniquely valuable—to their ecosystems, to science, and also to us."

—Beth Shapiro, author of *How to Clone a Mammoth*

"Matthew D. LaPlante is a rising star. In his new book, *Superlative*, he travels to the ends of the Earth to find the smallest, hardiest, most unusual organisms, and the interesting people who study them. As a professor of biology, I was shocked there was still so much I did not know about our brethren at the extremes. Hail evolution."

—David A. Sinclair, professor of genetics at Harvard Medical School

"*Superlative* is one of the very best books that I have seen in quite a while (and I read a lot). This is one of those rare books that you want to show people while going, 'Look at what it says here, did you know *that*?' LaPlante writes in an engaging and clear style that perfectly communicates his delight for nature's creativity while simultaneously lamenting the tragedy of extinction. As he describes astonishing examples of life's extremes, he seamlessly combines the sense of wonder that these examples inspire in him with a precise exposition of how this aesthetic knowledge can lead to practical applications, along the way articulating an ironclad case for the value of conservation efforts. *Superlative* should not only be in the library of any science enthusiast, but it should also be required reading for high school students and beginning college students. Highly recommended."

—Oné R. Pagán, PhD, professor of biology at West Chester University, and author of *Strange Survivors*

"*Superlative* displays a key scientific insight: It's the exceptions, the unusual, and the extremes that teach us the most. Matthew LaPlante's exploration of these exceptions is timely, fascinating, and exciting, giving us a chance to see what the future may—quite unexpectedly—offer us."

—Michael Fossel, author of *The Telomerase Revolution*

SUPERLATIVE

SUPERLATIVE

The Biology *of* Extremes

Matthew D. LaPlante

BenBella Books, Inc.
Dallas, TX

Although the African bush elephant, *L. africana*, is the largest land animal, my publisher much preferred the attractive foot of the Asian elephant for the cover. I concur. *E. maximus* might not be the biggest, but its feet are the prettiest.

BenBella Books, Inc.
10440 N. Central Expressway, Suite 800
Dallas, TX 75231
www.benbellabooks.com
Send feedback to feedback@benbellabooks.com

Printed in the United States of America
10 9 8 7 6 5 4 3 2 1

Library of Congress Cataloging-in-Publication Data:
Names: LaPlante, Matthew D., author.
Title: Superlative : the biology of extremes / Matthew D. LaPlante.
Description: Dallas, TX : BenBella Books, Inc., [2019] | Includes bibliographical references and index.
Identifiers: LCCN 2018048928 (print) | LCCN 2018050197 (ebook) | ISBN 9781948836210 (electronic) | ISBN 9781946885944 (trade cloth : alk. paper)
Subjects: LCSH: Animals--Anecdotes.
Classification: LCC QL791 (ebook) | LCC QL791 .L335 2019 (print) | DDC 591--dc23
LC record available at https://lccn.loc.gov/2018048928

Editing by Leah Wilson
Copyediting by James Fraleigh
Proofreading by Kimberly Broderick and Michael Fedison
Indexing by WordCo Indexing Services
Text design and composition by Katie Hollister
Cover design by Oceana Garceau
Printed by Lake Book Manufacturing

Distributed by Two Rivers Distribution, an Ingram brand
www.tworiversdistribution.com

To Heidi Joy

CONTENTS

INTRODUCTION: Nature's Best Ambassadors 1

CHAPTER I: All Things Great and Tall
How the World's Biggest Life-Forms Are Saving Human Lives 13

CHAPTER II: All the Small Things
Why Little Organisms Have Such a Big Impact on Our World 57

CHAPTER III: The Old Dominion
How Our Biological Elders Are Offering Us New Knowledge...... 95

CHAPTER IV: Fast Times
Why the Quickest Animals Probably Aren't the Ones You Think 131

Chapter V: Aural Sects
How Superlative Sound Helps Drive Life as We Know It.................... 159

Chapter VI: The Tough Get Going
How the World's Strongest Organisms Might Lift Us to the Heavens.................... 189

Chapter VII: Deadly Serious
Why the World's Most Efficient Killers Are Such Effective Lifesavers.................... 219

Chapter VIII: Smarter All the Time
Why the Most Intelligent Life-Forms Ain't Us.................... 259

Conclusion: The Next Superlative Discovery Is Yours 293

ACKNOWLEDGMENTS.................... 301
NOTES.................... 307
INDEX.................... 347
ABOUT THE AUTHOR.................... 373

Introduction

NATURE'S BEST AMBASSADORS

She was a month old and already heavier than me. But she was faster, too, and more energetic. And when Zuri the elephant bolted from her mother's side, kicking up dust and hay, and losing her footing a few times before reaching her tiny trunk toward my camera, I couldn't help but coo with delight.

This was exactly what I'd wanted, after all.

I had been covering national security issues for a newspaper in Salt Lake City for several years at the height of the wars in Afghanistan and Iraq. I'd made three trips overseas to write about the latter conflict. I'd witnessed death, despair, and desolation, and back home I was swimming in news about fraud, abuse, hopelessness, and ineptitude. There was the rash of suicides at a local military facility. There were the service members sickened by working at a former Superfund site. And there was military funeral after military funeral after military funeral.

I was sad and angry. All the time. And I needed to do something about it.

"Once in a while," I asked my editor, "do you think I could write about something happier?"

"Something like what?" he asked.

"Something like baby animals," I replied.

"Get the hell out of my office," he said.

I tried again the next day, and the day after. Eventually I won him over. So long as I kept up with my existing workload, he agreed, I could cover the local zoo, too.

A year later I found myself face-to-face with an infant elephant.

"Hey there, little one," I said, summoning the same lilting voice I used with my own baby daughter. "Welcome to the world."

Zuri paused and posed for a few seconds. She stretched out her big gray ears and tilted her head, then bolted back to her place beside her mother's sagging gray belly. A few moments later, she was back at it, slipping and somersaulting in the mud, and rolling about in the hay.[1] I was smitten.

But then, I've always been enamored with elephants.

I can still recall my first trip to a zoo. Or rather, I can recall the elephants. One, in particular, was an African bull named Smokey, whose enormity was intensified by the tragic smallness of his concrete pen. I stood in awe of him, feeling at once exultant and terrified. How could anything so big be real? I dreamt of him for years to come. I drew pictures of him in my school notebooks. I wrote stories about him in my journal.

I'm certainly not the only child to be obsessed by something superlative. Children are naturally enchanted by extreme things; if you doubt this, just try to wrestle a copy of *The Guinness Book of World Records* from the hands of an elementary school student. I got my first edition of that book in the second grade. I ordered it from the Warwick Elementary School book club, read through it in a matter of days, and then went back to the beginning and started again. When my daughter was eight, she spent hours leafing through the version of the book that now sits on my desk.

Of course she did. It's endlessly fascinating stuff.

Where is the world's oldest rainforest?[2] What insect produces the most painful sting?[3] Which dinosaur had the longest tail?[4] It's all in there, along with a bunch of far-less-natural records, like that held by a guy named Josef Todtling, who in 2015 allowed himself to be lit on fire and dragged 500 meters behind a horse for the dubious honor of saying no one in the known history of humankind had been pulled farther while ablaze.

Why do we do things like this? Why do we even care? Other animals don't even seem to notice when they encounter the biggest, fastest, strongest, or whatever-est of this quality or that. But we do. Almost as though it has been hardwired into us.

And maybe it was.

Cave drawings from around the world depict many of the largest animals our ancient forebears encountered, including mammoths, giraffes, bison, and bears. Our most ancient stories, and many of our newest ones, too, are replete with tales of tremendous giants, massive dragons, and colossal sea creatures—from Pangu, the giant horned deity who created the universe in ancient Chinese myths, to the many iterations of the gargantuan ape, King Kong, in modern American cinema.

Why has "biggest" always been such a big deal? Well, for ancient human hunters, the difference between killing an average antelope, gazelle, or wildebeest and killing the *biggest* of any of those creatures wasn't simply a matter of bragging rights—it was a few more days of food. As the "small-brained apemen"[5] who were our early hominoid ancestors struggled against a harsh and unforgiving world, the hunters who took down superlative game gave themselves and their families an edge on survival.

Not that bragging rights didn't matter. The cave folk who showed they were able to take down the fastest, strongest, or deadliest creatures in their neck of the woods were the best bets for mates who wanted to survive. That's a powerful selective force.

But while a fascination with extreme things seems intrinsic to who we are, science has often presented a rather adamant disinterest.

One example of this is the goliath frog of Middle Africa, which has been known to feast on turtles, birds, and bats.[6] In Cameroon they call the frog "bebe," because it's the size of a human infant.

Owing to their status as indicator animals, ones that are particularly sensitive to changing climates, frogs in general are a popular subject of research. Want to know how to quickly and accurately pick out the call of one species of frog from a cacophony of dozens of others? You can learn that and more in a study called "Acoustic classification of Australian frogs based on enhanced features and machine learning algorithms."[7] Curious about how frog health is impacted by water tainted with metals like copper, mercury, lead, and zinc? That's in "Antioxidative responses of the tissues of two wild populations of Pelophylax klepton esculentus frogs to heavy metal pollution."[8] All told, the ScienceDirect database of scientific, technical, and health publications includes more than 114,000 research articles about frogs.

How many of those publications so much as mention the world's largest frog? Nineteen as of the moment I wrote this—and just one of those studies, an examination of the parasites in the goliath's gut, is *specifically* about the goliath. In most of the other articles, the world's biggest frog gets a passing mention—for nothing more or less than being the world's biggest frog.

Upon detecting the first enchanting wisps of a novel research subject, many scientists are stricken with a bit of paranoia by the prospect of being scooped by another scientist. But when Claude Miaud set out to study the world's biggest frog, he wasn't worried about losing his place in line to another researcher. He was only worried about getting to the goliaths in time.

Miaud, one of the world's foremost frog experts and one of the only scientists to have spent much time with the goliath in its native habitat, believes the world's largest frog could very likely be pushed to extinction before any meaningful research has been conducted on it. At the time we began chatting about the goliath, in 2017, it had been more than a decade since anyone had so much as done an assessment of the frog's population in the wild. Meanwhile, the goliaths, which only exist in low-to-medium altitudes in southwestern Cameroon and northern Equatorial Guinea, have been widely hunted for food, and are suffering under the pressures of agriculture, logging, human settlement, and the sedimentation of the frogs' breeding streams.[9]

So while it was officially listed as "endangered" by the International Union for Conservation of Nature and Natural Resources (IUCN) in 2004, today it is likely *Conraua goliath* would garner "critically endangered" status—one precarious step above total elimination in the wild.

Extinction wouldn't just be a tragedy for the goliath. Knowing what an order of animals is capable of in the extreme helps us better understand the order as a whole. Yet from egg and larval development to calling, mating, and spawning behaviors, and even basic information like how long it lives, there's a lot we still don't know about the goliath—and might never get to know. "There are many black boxes in the knowledge of the biology of this species," Miaud told me. "Losing this species is limiting our current, and already limited, knowledge of biological phenomena."

Nonetheless, we have done very little to save, let alone study, the world's largest frog—which Miaud believes might be one of the longest-lived as well. If it is both very big and very old, that wouldn't be a matter of coincidence; organisms superlative in one way are often superlative in another, offering additional opportunities to witness, and learn from, biology in the extreme.

But without even the most basic information, like average lifespans, it's impossible to hazard a guess as to the goliath's processes of biological aging and cellular senescence, the halt of cellular growth. "Knowing the mechanism of senescence in wild animals," Miaud said, "is pertinent to understanding human aging biology."

That's not just speculative. As far back as the 1960s, researchers had noticed that the African clawed frog had a remarkably slow rate of aging,[10] and some of the first studies demonstrating a link between the length of telomeres (chromosomal "caps" that protect against deterioration) and aging were conducted on that same species, also known as *Xenopus laevis*. Today, the role of telomeres in human aging is a significant focus of many researchers engaged in the battle against age-related illnesses like Alzheimer's, osteoporosis, and cardiac disease.

Some people construct model railroads. Others paint with water colors. One of my hobbies is introducing scientists from different disciplines to one another.[11] I recently arranged a conversation between herpetologist Grace DiRenzo, who studies frog populations in Central America, and biochemist Laura Niedernhofer, one of the world's leading experts on the biology of aging. In a matter of minutes, the two scientists were geeking out about what frogs might be able to tell us about aging.

Because frogs appear to experience senescence quite slowly, it's not all that easy to find an aged frog in the wild. "I've never seen an old frog," DiRenzo told us. "I can't even tell you what an old frog looks like, and I've been in the jungles for a while."

Niedernhofer was instantly fascinated. She had just returned from a conference on aging at Xavier University, where she gave a presentation on mice, which age in similar ways to humans, right before Andrea Bodnar from the Gloucester Marine Genomics Institute gave a presentation on sea urchins, which don't appear to age at all.

"We get a lot of information on why we age by looking between species," she said. Mice and sea urchins, though, are separated by a whole lot of evolution. Mice and frogs? That's a closer genetic match, allowing for an easier hunt for genes that might impact aging.

If frogs do become a model organism for aging, as some researchers have suggested they should be,[12] what could the world's oldest frog tell us? We might never know. And if we lose the goliath, Miaud said, we'll lose a window into ourselves.

Consider the irony. There are few creatures humans have studied as much as those belonging to the order Anura. Hundreds of thousands of frogs are killed each year as part of a high school rite of passage intended to teach students musculoskeletal structure, cardiac physiology, and neurophysiology.

The frogs we study the most, like those from the genera *Xenopus*, *Bufo*, and *Rana*, exist right in the middle of a size spectrum that has the house-cat-sized *Conraua goliath* on one side and the pencil-tip-eraser-sized *Paedophryne amauensis* on the other.[13] Our favorite frogs for research are right in the middle of the weight spectrum. The amount of food they need is in the middle of the metabolism spectrum. And so on. Pick a trait all frogs have, and chances are good the frogs we use for research are pretty darn average when it comes to that trait.

They also tend to be very easy to find in the natural world. One of the most frequently studied frogs, *Rana temporaria*, can be found all over Europe. Its common name is actually "the common frog."

It's not just the goliath frog that has been widely ignored by researchers. The world's largest earthworm, the 10-foot Gippsland earthworm of Australia, is the titular subject in just three papers cataloged by ScienceDirect. The mammal with the smallest brain, the pygmy mouse lemur of Madagascar, isn't the main subject of any article in that database.

In one way, science's lack of focus on superlative organisms makes sense. Extreme phenomena are, by definition, outliers. And science has a very rational bias against data existing at the furthest, thinnest edges of the bell curve, which are often excluded to stabilize extreme variability that can cloud statistical analysis.

The predisposition to brush off outliers has also been justified by the sort of straight-up utilitarianism that guides much research. Penicillin has saved innumerable lives since it was discovered in the 1920s, but it has also taken a few lives along the way. That's because some folks—like those who carry a variant in the gene for interleukin, a protein that stimulates the immune system—may have a significantly increased risk of an allergic reaction to penicillin, leading to rashes, fever, swelling, shortness of breath, anaphylaxis, and yes, in some very rare cases, death.[14] For those people, Alexander Fleming's gift to humanity isn't much of a gift at all. But if doctors declined to prescribe antibiotics to anyone based on the idea they don't work for everyone, a whole lot of us wouldn't be here right now.[15]

Over the past few years, however, a lot of scientists have begun to recognize the long-sidelined potential of superlative organisms. There has been an explosion of scientific and conservationist activity around biological extremes, especially plants and animals at risk of extinction—organisms the Zoological Society of London calls "evolutionarily distinct and globally endangered," or EDGE species. Since 2007, the society has been sending research fellows around the world to study animals like the Chinese giant salamander, which is the world's largest amphibian; the olm, which can go longer without eating than any other known vertebrate; and the three-toed sloth, which has the lowest known metabolic rate of any mammal in the world.

These organisms aren't just biological Rosetta stones. They're creatures people can't help but pay attention to. That makes them great scientific ambassadors. And science is in dire need of ambassadors. For although it doesn't always seem so, the broad movement of history is toward an increasingly democratized world, where leaders must make choices based, at least in ample part, on what they believe their constituents want. In such a world, a broad lack of familiarity with something—even something objectively important—can spell its doom.[16]

At Utah State University, where I teach news reporting, feature writing, and crisis journalism, I often tell my students that to get readers to pay attention to anything important, you have to make it interesting. Extreme organisms are unquestionably interesting, and that allows us to tell important stories about ecology, conservation, research, and scientific history to people who might not otherwise think they're all that interested in science.

Fortunately, the ambassadors are all around us. There is no place in the world where, if you look hard enough, you cannot find an organism that is superlative in some way. You can find them living in the moss growing on the fence in your backyard, hanging out in the trees at the nearest park, and even darting about on the sidewalks near your home.

These life-forms don't just help us understand organisms closely related to themselves. They can help us understand our very place in this universe.

Want to understand how big we are? It's helpful to take a step back from biology, and wrap your mind around the idea of neutrinos, the smallest quantity of matter humans have ever dared to imagine—so small, in fact, that billions of them, moving at nearly the speed of light, are passing right through your body at this very moment.

Want to understand how small we are? It helps to know the observable universe is currently 93 billion light-years from end to end. To put this into context, the Voyager 1 space probe, which left our planet in 1977 and is moving away from us at about 40,000 miles per hour, is now approaching mile marker 14 billion on the cosmic highway to a region of space occupied by a star called Gliese 445. Assuming it avoids any interstellar catastrophes along the way, it will arrive in 40,000 years—at which point it will not have moved much at all in relation to our quickly expanding universe.[17]

When we think about things in this way, it accentuates our connection to every other living thing on planet Earth. It's all constructed from the same tiny collection of 118 known elements. It all exists in this infinitesimally small place in time in the 4.5-billion-year history of this planet, "a speck, orbiting a speck, among still other specks"[18] in our incomprehensibly big universe. And its genetic code is spelled out with the same four nucleic acids—often in long, shared sequences.

It's precisely because everything we know on this planet is so similar that it's so worthwhile to examine the stuff that, at a glance, seems so different from us and from others of its kind. The superlative phenomena of our natural world—from the fastest animals to the smallest life-forms to the slowest-evolving creatures to the oldest organisms—give us perspective on our parameters. That, in turn, helps us understand our potential.

That potential has been right in front of us for a very long time, not-so-well-hidden in the biggest, smallest, oldest, fastest, loudest, deadliest, toughest, and smartest things around us—things we've always been curious about, things we've always collected, things we've always told stories about.

I'm no longer a daily newspaper reporter, but I still do a considerable amount of journalistic work on some rather gloomy subjects. Ritual infanticide in Northeast Africa. The legacy of genocide in Southeast Asia. The deadly toll of gang warfare in Central America.[19] And I still try to balance that misery by doing things that bring me joy and awe.

To that end, a few years back, I began taking short excursions during my travels to get a glimpse of superlative life-forms. The world's biggest plants. The ocean's smartest inhabitants. The planet's deadliest predators. I didn't always find them. But I did find a lot of scientists who were eager to help me understand these organisms. It didn't take long for me to learn that the researchers who study these things often do so because they share my feelings of delight and amazement. By and large, they love talking about their work. And I love talking to people who love talking about their work.

I still visit Zuri, too. She's a lot bigger these days,[20] and a little less rambunctious, but we get together for lunch every now and then. I sit by her pen, she strolls over, and we have a chat.

I do most of the talking.

A few years ago, I dropped by with a large bandage covering the right side of my face. As Zuri sauntered over she seemed to stare a little longer at me than usual.

"Cancer," I told her. "No worries. It's all gone. The doctor cut it out and now I'm going to have a cool scar."

Zuri tilted her head and raised her trunk.

"I know, I know," I said. "Cancer isn't a problem you'll ever have, is it?"

She looked at me even more quizzically.

"OK," I said. "Let me try to explain."

Chapter I

ALL THINGS GREAT AND TALL

How the World's Biggest Life-Forms Are Saving Human Lives

Mago National Park is an astoundingly beautiful place, where morning fog gives way to spectacular vistas, and acacia trees are playgrounds for birds and baboons, the occasional python, and a small group of park rangers who have, in recent years, been losing a war against elephant poachers in southwest Ethiopia.

In the fall of 2017 Mago's game warden, who had asked park rangers for a census of the elephants in the park, invited me to join a foot patrol with a ranger named Kere Ayke, a member of the Kara tribe of the South Omo River Valley. Our hike began before dawn at a campsite along a rapid, muddy river, and twined through the mountains and valleys of the northeast quadrant of the park. But even before we'd left, Ayke was laughing off the task as rather pointless.

"There are elephants," he told me as we waded through the thick and thorny bush. "We know because we still see their tracks sometimes. But we haven't seen them in years. None of us have."

Sure enough, on that day we saw kudu, waterbuck, dik-dik, baboons, and guinea fowl. At one point something big and dark and gray seemed to slink into the bush 100 yards ahead of us or so, causing the treetops to tremble, but by the time we caught up it was gone, a ghost in the forest. We walked for miles, then miles more, on a route where the world's largest animals used to roam freely, and saw nothing more than trampled grass and hubcap-sized holes in the mud.

There are places in this world—some not far south from Mago even—in which elephants live in relative safety and where their numbers are stable or even growing. But that isn't the case in South Omo.

"The ones who are left, the survivors, are hiding," Ayke said. "They are terrified."

Back at the ranger station, conservation officer Demelash Delelegn told me why. Pulling an old ledger from the drawer of his rickety desk, and opening its yellowing pages to a hand-drawn spreadsheet, Delelegn released a sigh that seemed much too big for his slight frame.

"This is my very own handwriting from 1997," he said. "You see here? We found almost 200 elephants in Mago then."

That's how many they physically saw, Delelegn told me. Working from those numbers, the Ethiopian Wildlife and Natural History Society estimated there could have been as many as 575 elephants in Mago at that time.[1] By 2014 the estimated number, based on new surveys, had fallen to about 170. Still, Delelegn said, even a population that small, if properly protected, could grow.

"But it wasn't protected," he said. "We didn't do anything at all."

"So how many are there now?" I asked.

Delelegn's head dropped. "Not many," he said, tears welling in his eyes. "Not many at all."

He couldn't bring himself to even hazard a guess at a specific number. It was just too hard.

Outside his office, over a game of bottlecap checkers, I asked Ayke and his fellow scouts how many elephants they thought the most recent census would find.

"Maybe fifteen?" Ayke guessed. The other scouts nodded.

"There were 170 just a few years ago and now there are fifteen?"

They nodded again.

Mago is surrounded on all sides by tribes including the Kara, Hamar, Mursi, and Ari. The warriors from these tribes have always hunted elephants, though not in tremendous numbers. One elephant goes a long way, after all.

But when the government sold off tribal lands and permitted foreign investors to open factories in the region, some members of the tribes rebelled by doing something they'd been told they were no longer allowed to do: They began killing more elephants, trading the ivory for cash and guns.

Delelegn was surprisingly sympathetic. The scouts with whom he works are all drawn from the surrounding tribes. He said the government came and made demands of the tribes and didn't think of "how to talk to them in right and respectful ways."

The scouts are helpless to protect the animals. "We have forty-two people to patrol this entire park, which is more than 2,000 square kilometers," Delelegn said. "We don't have any trucks. We have very few guns. We have no education or training for this."

None of this came as a surprise to Nisha Owen, the manager of the Zoological Society of London's EDGE species program. There aren't enough game officers in the world to defend the species that need defending, she said. "Ultimately, protecting animals comes down to communicating with people," she said. "If we do it right we can have a tremendous positive impact, but if we do it wrong we can make things worse."

To do it right, though, requires understanding. And we got a very late start on that. It wasn't until 1977, in fact, that anyone thought to bring together a society of researchers around the scientific exploration of African and Asian elephants.[2]

A review of that community's work, in the first edition of the journal *Elephant*, shows just how poorly developed the field was in the 1970s. One zoo reported to the society that it was interested in the interbreeding of two separate species, an African bull and Asian cow—a futile biological enterprise and, today, a conservation anathema.[3] A list of "selected recent literature" included references to some more sophisticated scientific explorations, but even that list was thematically dominated by questions related to basic attributes, such as how long elephants live, how much they eat, and where they dwell—all things scientists could have explored decades or even hundreds of years earlier.

Fifteen years after the journal was born, its editor, biologist Jeheskel Shoshani,[4] was still lamenting just how much very basic information we had yet to gather about elephants. "We have just begun to understand elephant behavior and its role in the ecosystem," he wrote in *Elephants: Majestic Creatures of the Wild*, a thoughtful examination of the social, economic, and ecological roles of the world's largest land animal.[5]

Writing in the foreword to the book, the eminent zoologist Richard

Laws marveled at a then-recent finding that elephants communicate over long distances by ultra-low frequencies called infrasound.

It's not that such a discovery couldn't have been made during those years; infrasound had, at that point, been detectable by human instrumentation for seventy-five years.[6] But previously, no one had thought to even try. "There is no doubt," Laws added, "that there are further significant discoveries to be made."

Once we really started studying elephants, something became very clear: That was an understatement of gargantuan proportions.

HOW ELEPHANTS ARE LIKE MARTIAL ARTISTS

The African elephant is not so much an outlier as an outcome.

The largest living land mammal could be the poster creature for something called Cope's Rule. That nineteenth-century postulate, named for paleontologist Edward Cope,[7] suggests animals in a lineage tend to get bigger over time.[8]

From a long-term perspective, that's obvious. When life first appeared on this planet, some 4 billion years ago, it arrived in the form of single-celled organisms. By about 230 million years ago, those tiny creatures had evolved into dinosaurs, which ruled the planet for 165 million years to come, gradually getting bigger and bigger until . . .

. . . BOOM . . .

. . . an asteroid delivered a calamitous do-over, leaving very few large creatures on our planet's surface. The survivors were generally smaller burrowers that were able to wait out the upheaval. And, with those little animals as a launching point, Mother Nature started over again, rebuilding her creations, bigger and bigger with each passing millennium. Today our planet is once again walked upon by monsters. Giraffes and rhinoceroses. Hippopotamuses and elephants.

Cope's Rule isn't perfect. And within the vast fossil record it is certainly not hard to identify lineages that have reversed course in response to the conditions of the times and places in which they have lived. Over time, however, animals that don't die out tend to chunk up again. In this way, plotting the Cope effect would generally give us a chart that looks less like a straight line than an inverted roller coaster—up and down and up and down, but ever more up over time.

Like humans and every other placental mammal, elephants likely evolved from a furry-tailed, insect-eating, rat-sized animal resembling a long-nosed shrew and weighing in at about 8 ounces.[9] The following epochs gave us the 35-pound phosphatherium, which looked like a cross between a gerbil and a hippo. Later there was the 1,000-pound phiomia, which had a jutting jaw and thus looked more like a cross between an elephant and Jay Leno. Later still there was the 4,000-pound palaeomastodon, which looked like a modern elephant, but with a shorter trunk and tusks at all four corners of its mouth.

All of these creatures appear to be part of the lineage that gave us today's African bush elephant, the largest animal to roam the surface of our planet since the dinosaurs left Earth about 65 million years ago. (The largest of the mammoths rivaled their modern cousins in height, but likely fell short of the 15,000-pound weight of the largest African bush bulls.) Among an estimated 6.5 million other terrestrial species on our planet,[10] and the hundreds of millions of creatures that have lived on the Earth's surface since the last major extinction, the elephant is truly a fantastic beast.

The Darwinian explanation for how tiny animals turn into giant ones over time is fairly commonsensical; it is natural selection at work. Mutations producing a little more size and strength result in individuals better adapted to competing for mates, winning the battle for food, and fighting off predators.

But perhaps the single biggest driver of Cope's Rule is something so obvious it's often overlooked. As evolutionary biologist John Bonner once explained to me, "there's always room at the top." Smaller animals, he pointed out, must compete with a lot of other animals in a similar niche. But once an animal finds itself at the top of the heap within its biome, Bonner said, "they escape the competition." Thus, over time, a lot of animals get bigger.

And not just bigger, but different. For as a creature grows, the universal laws governing matter and energy begin to push and pull upon the forces that maintain life. In his book *Why Size Matters*, Bonner argues size is the driving force for all of biology, including evolution. It's why elephants look like elephants, rather than really big versions of our common ancestor. They didn't grow bigger because mutations like long snouts made them more fit, Bonner argues. Rather, such mutations were needed to accommodate their growth.

To explain how this works, Bonner likes to invoke the Brobdingnagians of Jonathan Swift's *Travels into Several Remote Nations of the World. In Four Parts. By Lemuel Gulliver, First a Surgeon, and then a Captain of Several Ships* (which you likely know as *Gulliver's Travels*). Like Swift's Lilliputians, who looked precisely like humans but were much smaller, the Brobdingnagians are described as looking just like humans, but standing twelve times taller than we are, or nearly 70 feet tall. But a creature that tall poses a problem of physics, because an increase in height would correspond to an increase in width, which would correspond to an increase in depth. And the area within those dimensions, filled to the brim with tissue and bone, Bonner estimates, would weigh something on the order of 12 or 13 tons.

Supporting that weight wouldn't be possible on legs simply scaled up from human dimensions. To have any hope of walking, the Brobdingnagians would have to have greatly thickened lower limbs, which

Bonner says would give them the appearance of being "victims of advanced elephantiasis of the legs."[11] It's not just that they wouldn't look human—they *couldn't* look human.

Elephants don't simply offer us an opportunity to understand the colluding forces of physics and evolution, though. They also have much to teach us when it comes to survival. Because even though they have a lot going for them in that respect, as the most extreme land-dwelling example of Cope's Rule, they also have a lot going against them.

While evolution seems to drive many creatures, and particularly mammals, to get bigger over time,[12] it also eventually blows the biggest animals off the extinctionary edge—a phenomenon I've come to think of as "Cope's Cliff." Bigger animals take a lot more food and water to survive from day to day, and are thus more prone to starvation in times of scarcity. They have longer gestation periods and tend to birth single offspring at a time, meaning they don't replace themselves, let alone grow their numbers, as quickly as smaller animals. As a result, they can't evolve as quickly in response to changing climates.

It is for reasons like these that we are no longer sharing the Earth with animals like *Castoroides*, a genus of goliath prehistoric beavers that could reach 7 feet long and weighed more than 250 pounds. *Castoroides* likely bowed out of existence around 11,000 years ago, along with the genera *Glyptodon*, an armadillo the size of a Volkswagen Beetle, and *Megalonyx*, a 10-foot-tall sloth. Ahead of all of those genera on the great and lethal leap off Cope's Cliff was the real Bigfoot—*Gigantopithecus*, a 10-foot and 1,000-pound fruit-eating ape that lived in what is now southern China.[13]

All huge. All gone now.

So bigger is better—until it isn't. That's what makes the world's largest land animal so special. Elephants have managed to survive both because of and despite their size. Somehow, they've successfully balanced upon an evolutionary tightrope strung between their sheer enormity and pressures like cataclysmic environmental changes, hungry predators, and the necessity to evolve. Having come to the precarious edge of Cope's Cliff, they've almost miraculously managed to keep from falling off.

And although elephants are the largest of all extant land animals, every life-form having achieved superlative size among its own lineage has walked a similarly incredible evolutionary path. Giraffes. Giant redwoods. Blue whales. These organisms are evolutionary jujitsu masters—the perfect combination of strength, balance, and fitness. Their size is a clue to us that they have tremendous knowledge to impart about surviving in this world, if we'd only just be willing to study at their enormous feet.[14]

WHY ELEPHANT CELLS ARE LIKE EMPATHETIC ZOMBIES

At first phenotypic blush, my friend Zuri and I don't seem much alike. She has a trunk and I've got thumbs. She has crackly gray hide and I've got freckled peach skin. She can communicate over long distances via low-pitched rumbles detectable by vibrations in her trunk and feet, and I've got text messaging, email, and Twitter. And, not for nothing, there's the whole size thing: She weighed far more as a 251-pound infant than I've ever weighed as an adult.

Physiologically speaking, it might appear as though you couldn't find another mammal more different from *Homo sapiens* than *Loxodonta africana*.

But there's also a tremendous amount of common ground. Both of our species live unusually long lives for mammals—decades after our last reproduction, a point at which a lot of other animals tend to knock off. We're both social creatures who live in highly complex communities. We both have relatively big brains, even when adjusted for body size.

And in our genomes? There's a lot of overlap. About three-quarters of elephant genes have human analogs—and hidden inside all of that shared genetic material is potential for deep understanding.

The genetic coding we share with elephants has been stressed and stretched in all sorts of different ways since our species diverged, but vast sequences in our DNA still perform markedly similar functions. So when we see something an elephant is doing with its genome, it's not beyond reason that—given the right circumstances—we could do it with ours, too.

One example of this came by way of the legendary biological anthropologist Morris Goodman, a pioneer in the use of DNA to chart the course of evolution. Goodman wanted to know whether the human solution to addressing the needs of big, oxygen-ravenous brains was, in fact, uniquely human at all. After sorting through the genomes of fifteen animals, all of which diverged from one another at vastly different times in the past 310 million years,[15] Goodman realized humans and elephants had both undergone an accelerated period of evolution in our "aerobic energy metabolism" genes, which impact how our mitochondria use oxygen.[16]

Here's the crazy thing: That rapid period of change happened long after humans and elephants broke ranks on the evolutionary tree of life—resulting in a shared genetic trait that evolved separately, and at different times in history, as our two species confronted the common challenge of growing big brains.[17] So it's not just the traits we shared at the time we diverged that are important—it's the potential within our genomes for those traits to develop when circumstances require it.

That potential is at the center of the burgeoning field of comparative genomics, and at the heart of the work of pediatric oncologist Josh

Schiffman, a childhood cancer survivor who is growing more convinced, every day, that he might be unlocking the secret to curing cancer.

It all started in the summer of 2012, when Schiffman's beloved dog, Rhody, passed away to histiocytosis, a condition that attacks the cells of skin and connective tissue. "It was the only time my wife has ever seen me cry," he told me. "Rhody was like our first child."

Schiffman had heard dogs like his had an elevated risk of cancer, but it wasn't until after Rhody's death that he learned just how elevated it was. Bernese mountain dogs who live to the age of ten have a 50 percent risk of dying from cancer.

"Suddenly it dawned on me there was this whole other world, this young field of comparative oncology," he said, "and I was pulled into the idea of being a pioneer and maybe a leader to help move things along."

Schiffman had long been intrigued by the fact that size doesn't appear to correlate to cancer rates—a phenomenon known as "Peto's Paradox," named for Oxford University epidemiologist Richard Peto. But when Schiffman took his children on an outing to Utah's Hogle Zoo—the same place I sometimes go to have lunch with my elephant friend, Zuri—everything came together.

A keeper named Eric Peterson had just finished giving a talk to a crowd of visitors, mentioning in passing that the zoo's elephants have been trained to allow the veterinary staff to take small samples of blood from a vein behind their ears. As the crowd dispersed, an angular, excited man approached him.

"I've got a strange question," Schiffman said.

"We've heard them all," Peterson replied.

"OK then—how do I get me some of that elephant blood?" Schiffman asked.

Peterson contemplated calling security. Instead, after a bit of explanation from Schiffman, the zookeeper told the inquisitive doctor he'd look into it. Two and a half months later, the zoo's institutional review board gave its blessing to Schiffman's request.

Things moved fast after that.

Cancer develops in part because cells divide. During each division the cells must make a copy of their DNA, and once in a while, for various reasons, those copies include a mistake. The more cells divide, the greater the odds of an error, and the more prone an error is to be duplicated again and again.

And elephant cells? Those things are dividing like crazy. Based on the number of cell divisions elephants need to get from Zuri's size when we met to the size she is now, in just a few short years, it stands to reason they should get *lots* of cancer. Yet they almost never do.

"Going from 300 pounds as a calf to more than 10,000 pounds, gaining three-plus pounds a day, they're growing so quickly, so big and so fast—baby elephants really shouldn't make it to adulthood," Schiffman said. "They should have 100 times the cancer. Just by chance alone, elephants should be dropping dead all over the place." Indeed, he said, they should probably die of cancer before they're even old enough to reproduce. "They should be extinct!"

Already, comparative oncologists suspected the exceptionally low rate of cancer in elephants had something to do with p53, a gene whose human analog is a known cancer suppressor. Most humans have one copy—two alleles—of the gene. Those with an inherited condition known as Li–Fraumeni syndrome, however, have just one allele—and a nearly 100 percent chance of getting cancer. The logical conclusion is more p53 alleles mean a better chance of staving off cancer. And elephants, it turns out, have twenty of them.

The big find that came from Schiffman's exploration of the elephant blood he got at the zoo, though, was not just that there were more of these genes in elephants, but that the genes behaved a little bit differently, too.

In humans, the gene's first approach for suppressing tumor growth is to try to repair faulty cells—the sort that cause cancer. So, at first, Schiffman's team assumed having more p53 genes meant elephants had bigger repair crews. With the goal of watching those crews in action, the researchers exposed the elephant cells to radiation, causing DNA damage. But they noticed that, instead of trying to fix what was broken, the elephant cells seemed to grow something of a conscience.

To understand this, it's helpful to think about how you'd respond in a zombie apocalypse. Of course you'd fight long and hard to keep from being infected, right? But if a zombie was about to chomp down on your arm, and there was nothing you could do to stop it, and if you had but one bullet remaining in your gun—and a few moments to consider what you might do to your fellow humans as a part of the legion of the undead—what would you do?

That's what elephant cells do, too. Under the directive of p53, mutated cells don't put up a fight. Upon recognizing the inevitability of malignant mutation, they take their own lives in a process known as apoptosis.

And they don't just do this for one kind of cancer. The p53 gene apparently programs cells to do this in response to all kinds of malignantly mutated cells in elephants—a finding that flies in the face of the conventional assumption that there is no one singular cure for the complex group of disorders we call cancer.[18]

When I first met Schiffman in 2016, he was brimming with excitement about the potential elephants have to help us understand cancer. He was also very cautious not to suggest he was anywhere near a cure, nor that he ever would be.

Just a few years later, though, Schiffman was speaking openly about his intention to rid the world of cancer. And, to that end, what's happening in his lab is encouraging, to say the least.

He and his team have been injecting cancer cells with a synthetic version of a p53 protein modeled on the DNA he's drawn from Zuri and other elephants from around the world. Viewed on time-lapse video, the results are unmistakable and amazing.

Breast cancer. Gone.

Bone cancer. Gone.

Lung cancer. Gone.

One by one, each type of cancer cell falls victim to zombie-cell hara-kiri, shriveling and then exploding, and leaving nothing behind to mutate.[19] Schiffman is now working with Avi Schroeder, an expert in nanomedical delivery systems at Technion-Israel Institute of Technology, to create tiny delivery vehicles to take the synthetic elephant protein into mammalian tumors.

If this was all the benefit we ever derived from studying elephants, it would be plenty.

But it's not. Not at all.

WHY ELEPHANT AROUSAL IS GOOD FOR OTHER ANIMALS

It was a summer day at the Oakland Zoo, one of very few accredited animal parks caring for bull elephants in the United States.

There were dozens of visitors milling about near the elephant exhibit that afternoon but, up to that point, few of them were really looking at the twenty-three-year-old beast named Osh—not until he ambled, huge head bobbing up and down, closer and closer, and came to rest in the corner of his pen nearest to where visitors could stand.

From there, the young bull's enormity was appreciable, breathtaking. At the time he stood 10 feet, 9 inches tall and weighed 13,000 pounds. His legs were as big around as many of the trees lining his pen. His eyes were gleaming amber and his ears were like beating aortas; as they flapped, they clapped against his shoulders and sent great clouds of dust into the air. His rumble had the cadence of a cooing pigeon and the octave of a distant train.

The visitors watched with giggles and sighs and oh-my-Gods as Osh lifted his trunk over the fence and peered into an adjoining pen where the female elephants were kept. But then, from the gallery, came the gasps of horrified parents and the quick shuffling of children's feet as they were turned and pulled away by their adult companions, as Osh's penis appeared between his back legs, his erection growing larger as he stared at the cows.

That Osh was clearly interested in his female companions was a good sign. It had been just over a year since his first musth, a rutting period when male elephants become uncharacteristically aggressive, and his keepers were cautious but eager to get him in a pen with the cows. Every fertile new bull, after all, offers an opportunity to add to the diversity of genes in the captive population, a collection of animals zoo advocates argue are vital for species conservation.

At that point, in 2017, artificial collection of Osh's semen had proven unproductive owing to urine contamination. Colleen Kinzley, the zoo's animal care director, told me she was hoping a more "let nature take its course" approach might change the reproductive dynamics, but Osh needed to take the lead on that.[20]

The challenges Kinzley and her staff were facing with Osh aren't unusual. As wild populations of elephants continue to plummet, the Association of Zoos & Aquariums has pressed its members to make breeding a priority, but captive breeding of just about any wild species is replete with challenges. Overcoming those challenges takes creativity, patience, luck, experience—and money.

We might have gotten a late start on studying them, but these days hundreds of scientific reports are published each year on wild and captive elephants, with many of the studies focused on discoveries about breeding and reproduction. Among the leaders in this field is Wendy Kiso, who in the mid-2010s was a researcher at the Ringling Bros. Center for Elephant Conservation in Polk City, Florida, where dozens of former performing pachyderms were sent after the circus shut down its elephant act. Among other findings, Kiso has discovered novel ways to help ensure elephant sperm remains viable after freezing. Diluting the semen in an egg yolk–based solution, it turns out, helps in this regard.[21] Kiso and her team have also discovered that variability in the quality of sperm in Ringling's elephant herd can be predicted by an elephant's levels of lactotransferrin, an antibacterial protein found in all mammals that helps transfer iron into cells.

These sorts of studies can seem esoteric. Here's why they're not: Just as we can learn a lot from elephants when it comes to fighting cancer in humans, we can learn a lot about how to save other animals by what we now understand about preserving pachyderms. The research Kiso's team has conducted has been used again and again as the building blocks for scientists studying—and seeking to save—other animals, including those that don't command nearly as much affection, public fascination, or funding as elephants do.[22]

There are more than 40,000 species on the IUCN's Red List of Threatened Species. Thousands have been declared "critically endangered" and there are likely thousands of others that would garner that designation with an updated population assessment. Very few of these animals have captured the hearts and reached into the pocketbooks of people around the world in the way elephants have.

One of the key struggles in saving any endangered animal is preserving genetic diversity. That's why conservationists have been working on developing genome resource banks—repositories of sperm, embryos, tissue, blood, and DNA that serve as an "insurance policy" that there is enough genetic material, from as many different sources as possible, to give us the best chance of saving animals on the brink of extinction.

Much of what we know about preserving the sperm of endangered animals has come from what we've learned from the billions of dollars we've spent on research into human fertility, and from livestock heavily bred for reproductive success. If it works for *H. sapiens* and it works for *Bos taurus*, we've generally assumed it'll work for everything else. Kiso's team showed that wasn't the case. Even species as closely related as *Elephas maximus* and *L. africana* (which diverged about 7.5 million years ago) needed to be treated differently. Another of their studies, for instance, demonstrated the different ways the sperm of Asian and African elephants responds to various freezing techniques.[23] Up to that point, the sperm of the largest and second-largest land animals in the world was generally handled the same.

Two years after that paper was published, prominent biologist Pierre Comizzoli, the director of the influential Smithsonian Consortia, used the Ringling team's findings to help make the case for investment in reproductive research on other animals. Wild ones. Endangered ones. Ones that often have a lot more in common with fellow threatened species than people and cows do. Despite the fact that cryopreservation is a vital part of international conservation efforts, Comizzoli lamented, "virtually all other species . . . have gone unstudied."[24]

Now that's starting to change, thanks to the elephant's very big coattails. And that's not the only way in which animals that have captured our collective fancy with their immense size are changing long-held scientific assumptions.

WHY ALMOST EVERYTHING WE KNOW ABOUT GIRAFFES IS WRONG

It was the trademark example of Lamarckian inheritance—the theory, pushed forward in 1801 by Jean-Baptiste Lamarck, that more frequent and continuous use of any specific characteristic would gradually strengthen that characteristic, pushing an animal toward "the limit of its development."[25]

"In places where the soil is nearly always arid and barren, it is obliged to browse on the leaves of trees and to make constant efforts to reach them," Lamarck wrote in 1809 of giraffes and their exceptionally long necks. "It has resulted that the animal's fore-legs have become longer than its hind legs, and that its neck is lengthened."

Many of the ideas Lamarck espoused were supplanted as Charles Darwin's evolutionary theories took hold following the publication of *On the Origin of Species* in 1859, and much of what did remain en vogue about "use and disuse inheritance" was abandoned when Mendelian inheritance became the core of genetic theory in the early 1900s.[26] But the giraffes-got-tall-to-reach-the-leaves idea stuck around—with Darwin himself giving the hypothesis an added boost. Giraffes, he wrote, were "beautifully adapted for browsing on the higher branches of trees."[27]

These days, even if they don't know the difference between Lamarckism and Darwinism, most people will tell you that giraffes, however they evolved to be the tallest animal in the world, did so in order to reach the leaves on tall trees.

There's just one problem with that theory: There isn't much evidence to support it.

Back in 1991, Lynne Isbell, a veteran field researcher who is something of a specialist at making observations other scientists have missed, was among the first people to note that giraffes don't often use their full height to feed. Most of the time, in fact, they bend *down* to eat.[28] That's an observation that was later confirmed by other scientists, who noted that in the dry seasons—when feeding competition should be most intense and thus offer the greatest selective pressure—giraffes are even more likely to eat from the same low shrubs as competing browsers.[29]

Isbell said the A-causes-B way we tend to think about evolution is, quite simply, too simplistic. "One of the things I love is selective pressures," she told me, stressing the "s" in "pressures" to emphasize its plurality. "The environments in which animals live and in which they evolve are exceptionally complicated, so there are always multiple selective pressures going on at once."

While it's definitely possible that leaf-reaching was one of many factors that led giraffes to their superlative status, scientists like Isbell have offered a lot of compelling evidence that it wasn't the main factor, let alone the only one, as so many people have been taught.

Giraffes don't just use their long necks and legs to eat, after all. They also use their necks to fight. Males use their thick skulls as battering rams; the longer and heavier the neck, the more advantage a bull has when competing with rivals for cows. And they use those long legs to kick and run, two key survival skills for the fight-or-flight realities of life in the midst of some of the most ruthless carnivores on the planet.

Rather than being the product of simplistic cause-and-effect evolution, giraffes' superlative height is more likely the result of a perfect storm of selective pressures—the ability to reach food, yes, but also the competition for sex, the need to outrun predators, and the ability to stand and fight when necessary to protect themselves and their calves. All of this would help explain why, when scientists at Penn State University compared several giraffe genomes to the genomes of the okapi—a much-shorter giraffe cousin whose hindquarters resemble a zebra—they found seventy different genes that had undergone significant changes in the relatively short 11 million years since the two lineages diverged.[30]

Genetic sequencing didn't just lend evidence to the idea that giraffes evolved to be tall as the result of myriad selective pressures, though. It also put a hole in what now appears to be a foundational misunderstanding about the species—namely, that it's *a* species.

Axel Janke understands why people were reluctant to believe him when he first suggested that we might have to change the way we think about giraffes. The pioneering geneticist said he didn't even believe it himself, at first, when the DNA samples he tested from giraffes throughout Africa showed a surprisingly vast genetic diversity.[31] "I said, 'Something is odd here,'" he recalled. "I didn't suspect different species at this point, but when we analyzed the data with different methods, you could see four groups."

Even then, he said, he was hesitant to suggest something so clearly contradictory to the conventional wisdom about the giraffe "because—come on—everybody knows giraffes," he said. "If you ask a four-year-old what a giraffe is, she can explain it to you."

"Yeah, it's the tall one," I said.

"Precisely," Janke said. "Everybody knows that. But not much else . . . and when we think of these animals based on one attribute, we tend to miss things."

The decision to classify giraffes as a single species came in the 1700s, and was made unilaterally by a guy who had never actually seen a live one and who either didn't realize or didn't much care that giraffes come with a tremendous variety of body shapes, patterns, behavioral characteristics, geographic ranges, and mating behaviors. That guy, though, was Carl Linnaeus, the father of biological taxonomy. And the figurative shelves he created for the 12,000 species of plants and animals he classified have proved to be quite sturdy, even sacred. Even though it's clearly an imperfect system, and substitutes have been proposed, we continue to use the binomial system of classification Linnaeus developed more than 250 years ago.[32]

Yet what Janke discovered, when he put Linnaeus's single-species assumption to the genetic test, is that various groups of giraffes are as different from one another as polar bears are from grizzly bears. He's identified four separate species, including the southern giraffe, the reticulated giraffe, the northern giraffe, and the Masai giraffe, the latter of which is the tallest of the four species and thus the tallest mammal in the world. Janke told me he thinks a genetic case might be made for other giraffe groups to be considered separate species as well, but that will take more time and research.

Now, if you're thinking, "But wait—can't these different groups of giraffes interbreed?" then you're not alone. One of the most common beliefs about speciation is that it's all about fertility. It's not.

The definition of species most scientists use, at least when it comes to organisms that reproduce sexually, comes from the taxonomist Ernst Mayr, who suggested that speciation comes from reproductive isolation.[33] While that isolation can certainly come from biological barriers, it can also come from geographic or behavioral barriers.

And the different groups of giraffes are indeed isolated. Very isolated. Janke said he knows of no hybrids in the wild and only one case in a zoo. Yes, two giraffes from different groups *can* potentially produce fertile offspring but, as the different giraffe populations evolved, they didn't—not for more than a million years. That's further back in time than the point at which most species of archaic humans, including *Homo neanderthalensis* and *Homo rhodesiensis*, split from our common ancestor.

All of this could fundamentally impact the way we approach giraffe conservation, because even though myriad factors affect whether an animal is declared to be endangered, sheer numbers are a big part of the calculus. As recently as 2010, the IUCN declared the giraffe to be a (single) species of "least concern" when it comes to the likely danger of extinction. When you split the 100,000 remaining wild giraffe into four separate species, though, things begin to look a lot more dire, especially since those groups aren't split equally. The northern giraffe, for instance, has just a few thousand individuals remaining in the wild, and those animals are dispersed in groups across thousands of miles of Africa. Old assumptions die hard, though, and as of 2018 the union still hadn't shifted its assessment that, as a singular species, the giraffe was vulnerable, but not endangered.[34]

The genetic evidence is compelling, so I think the union will change its tune—and soon. And, once that happens, there may be greater reason for hope for *all* of the different giraffes. Because the union's listings carry moral and legal weight in nations around the world.

And when we get together as an international community we can have a really big impact on really big things.

WHY BLUE WHALES ARE SO HARD TO RESEARCH

I saw the breach off the port bow, perhaps 200 meters to the east of the boat.

"Whale," I said, pointing through a foggy windshield. That was all I could manage, for I'd never seen anything like it.

I served in the US Navy when I was younger, and circumnavigated the globe on the USS *Nimitz*. But aircraft carriers are awfully big ships, and there were days and even weeks when I did not get topside, the hours marked by bells and duty changes instead of sunrises and sunsets. Occasionally, in the middle of the night, I'd sneak onto the aft deck and marvel at the bioluminescent trail our enormous ship left in its wake, a magical gift from some of the smallest inhabitants of the ocean, but I had never seen the biggest.

Twenty years later, on a much smaller vessel in California's Monterey Bay, I was making my first acquaintance with members of the family Balaenopteridae.

"Is it a humpback?" I asked my guide, marine biologist Nancy Black, as a huge gray tail suddenly appeared out of the water, slapping the surface, again and again, with such force we could hear it in her boat's cabin.

"It is," she said.

"It's enormous," I replied, still trying to manage my awe.

"It's just a little guy," she laughed. "Just a playful puppy."

Black's nonchalance was understandable, I suppose. She's been at this for thirty years. There are few people in the world who have seen more whales than she has. And this encounter followed, by just a week, a visit to that same bay by the biggest of the sea giants. One of Black's crew members was flying a drone on that day, and the overhead image of a blue whale, surfacing near a watercraft less than half its size, was something to behold.

As is the case for so many extreme creatures, humans got a pretty late start when it comes to scientific inquiry about the biggest of them all. Until the International Whaling Commission's 1986 ban on whaling—and even after that in some areas of the world—researchers had to compete with hunters to reach the blues. In this respect, we're fortunate whales weren't pushed off Cope's Cliff altogether. It's a testament to the evolutionary fitness of these masters of our oceans that our centuries-long effort to harpoon as many of them as possible didn't result in more extinctions.[35]

And when it comes to *Balaenoptera musculus*, in particular, that permits us to say something rather extraordinary: We exist on this planet at the same time as the largest animal to ever live.

To fully appreciate how special this is, do what Eric Kirby suggests doing: Take a walk on a football field. A 100-yard gridiron is a graspable stand-in for the 4.5-billion-year history of our planet. If you started on the northwest goal line of Reser Stadium, at Oregon State University, where Kirby teaches geology,[36] and started walking southeast from there, you'd be well within the 2-yard line on the opposite end of the field before you got to the place where our modern mountains were born, and standing right about on top of the 1-yard line when you got to the place when whales arrived. By the time you got to the place where blue whales came along, you'd be 3 inches away from the end zone. And within those few remaining inches, Kirby told NPR a few years ago, "if you pluck two hairs out and lay them down on the goal line, that's about how long we've had civilization on our planet."[37]

And yet there we are. And there they are, too. What luck.

Well, for us, at least. A century ago, there were hundreds of thousands of blue whales in our oceans. Today there are perhaps 25,000 left. Even as humpbacks and gray whales have recovered from a century of commercial whaling—a true testament to the effectiveness of international cooperation—the blues have lagged behind throughout the world.

There are exceptions, though—places where blue whales are thriving—and that's what prompted me to go looking for whales on the California coast. Some researchers believe blue whale numbers have rebounded in this area to near-historic levels—a phenomenon that does not appear to be the case with other populations of blues.[38] Something happening in California's coastal waters is bringing the world's largest animal—all 100 feet and 300,000 pounds of it—back from the brink.

What? That's not precisely clear. For despite their gargantuan size, blue whales are actually quite hard to find and even harder to research.

That might seem counterintuitive. It's hard to miss a blue whale, right? But as big as they are, the ocean is so much bigger, and sometimes we forget just how big it is.

If every remaining blue whale in the world gathered together in one area of the ocean's surface, lining up side by side and nose to tail, they could pack themselves into a few square miles. But the total surface of our ocean is nearly 140 million square miles, and the blue's territory encompasses most of those waters.

There are places where they're a bit easier to find, though—if we're willing to challenge assumptions. That's what Kirby's colleague at Oregon State, marine ecologist Leigh Torres, discovered in the South Taranaki Bight, a bay on the southwest coast of New Zealand's North Island, which is also known as Te Ika-a-Maui. Torres had heard stories about blues in the bay, and lots of Kiwis knew the world's largest creature could be spotted in the bight, but when Torres began to dig into the scientific literature in the early 2010s, she found no one had so much as confirmed the blues' presence in the bay. "It was a case of, 'Oh yeah, they show up there sometimes,'" she told me. "But no one knew much more."

Intrigued, Torres began digging into historic whaling records, studying oceanographic data and examining Taranaki's plant and animal life. "Everything seemed to point to the idea that the bight wasn't just a place that blue whales might be found, but that it was really an ideal environment for them," she said.

There were, at that point, only a few known foraging grounds for blues, places where scientists could go to have any hope of encountering enough whales to conduct reliable research. Vitally, these are often places where the whales get even more protection under international agreements and national laws—places where they could go to make baby blues.

In a 2014 expedition, Torres and her team set out to demonstrate the blues were in the bight. They accomplished that goal easily, and made another discovery in the process: The Taranaki blues were genetically distinct from other blues that frequent New Zealand. And when the team sent photos of more than 150 individual whales to researchers around the world, they learned something even more stunning: None of the animals had ever been identified anywhere else. By way of contrast, almost every other whale that has been identified around New Zealand has been spotted elsewhere in the world.[39]

Blue whales are typically thought to be the some of the Earth's greatest nomads, traveling thousands of miles each year in a never-ending quest for food and favorable breeding locations.[40] The whales Torres and her team identified, though, appeared to spend nearly all of their lives in Taranaki—a finding backed in more recent years by hydrophone recordings establishing the presence of blues in the bight throughout the year.

Thanks to Torres, researchers now have a place to go where they can be almost guaranteed to find the world's largest animal. And already this has offered us a wealth of new information about creatures that have previously eluded scientists.

One such finding came in 2017 when drone videos captured by Torres's research collaborator and husband, Todd Chandler, showed Taranaki blue whales feeding on huge patches of krill—and passing up opportunities for smaller bites. In one video, a blue locks onto a cloud of pink crustaceans, rotates sideways and opens its mouth, filling its ballooning gullet. Opening its mouth was like throwing open a parachute: The decision to feast slowed the animal from nearly 7 miles per hour to just 1 mile an hour, a shift that meant the whale would have to exert significant energy in order to bring itself back up to feeding speeds for the next mouthful.

In another video, the same animal begins an almost identical approach on another, smaller patch of krill, but appears to decide at the last moment not to open its mouth, apparently making an energy-versus-calories calculation to wait for a bigger meal.[41]

Torres believes blue whales make these choices all the time, and really quickly, based on a number of sensory inputs. Certainly, she said, vision plays a role, but so might smell, sound, and even the "feel" of water being disturbed by thousands of tiny organisms with their flailing antenna, legs, setae, and gills.

Humans—brilliant, brainy creatures we so often believe ourselves to be—are pretty bad at this sort of thing. Engaging one sense often comes at a cost to our other senses.[42] Whales, on the other hand, appear to have the ability to form a synergistic, rapid-fire-decision-inciting picture from various sensory inputs.[43]

And why wouldn't they? We know intelligence is a function of brain size, neocortical surface area, the number of neurons, and the impact of rapid evolution, among other factors. Whales have all of that in spades. Just as we have learned a lot about ourselves from studying the world's largest land animal, the world's largest sea animal has a lot to tell us about how our brains work.[44]

That's true, however, only if places like Taranaki continue to offer us the opportunity to observe these otherwise elusive creatures. And, unfortunately, right now there's no guarantee that safe harbor will continue to be safe.

In the summer of 2017, New Zealand's Environmental Protection Authority gave Trans-Tasman Resources, an underwater extraction company, approval to dig up 50 million tons of iron sand each year from the Taranaki seafloor—separating out the ore and dumping the rest back into the bight. The company had argued there's already plenty of commercial activity in the bight. While that's true, none of the current activities amount to turning the bay into a virtual snow globe of silt from the seafloor. The decision was immediately appealed to New Zealand's High Court.

Torres is hopeful her research will impact the bight's future. And it might. Because when we protect the biggest of things, we protect a whole lot of other creatures, too—including rare fish like the tarahiki and the conger eel. Threats against those sea creatures might not garner the same sort of outcry as the one now confronting Trans-Tasman thanks to the bight's blue whales. Even before the High Court got the case, nearly 14,000 Kiwis had submitted letters of protest.

"People love megafauna," Torres said. "And the more they learn, the more protective they are."

HOW WHALE POOP AND A TERROR ATTACK ARE HELPING US UNDERSTAND STRESS

We pushed off from the harbor on a Saturday morning, just as the final wisps of fog were clearing off the coast of Newport, Oregon. Two bald eagles watched us from the rocky jetty. The ocean was silken.

We hadn't been underway for more than ten minutes when we spotted Pancake. The frisky teenaged female, recognizable by a round white splotch on her side, was tracing long, graceful circles in the water just to the north of the harbor's mouth.

"This is the only place I've ever been where finding the whales isn't the hard part," Leigh Torres told me as Todd Chandler maneuvered our bright orange Zodiac boat into position behind the 40-foot gray whale. "You just leave the harbor and there they are."

The grays here are a fascinating lot. They're part of a group of about 20,000 that each year starts a migration in their winter breeding grounds in Baja California and then heads north to their summer feeding grounds in the Bering Sea.

Pancake is among about 200 members of that group, though, who don't make the full trip. They pull up short—really short—along the central Oregon coast, where they wait for the rest of the crowd to head up to Alaska and head back. Give or take a few newborns or recently departed elders, it's the same whales every year. Pancake, for instance, has been spotted in these waters since 2002.

Maybe they're lazy. Maybe they're smart. Nobody's yet sure, in no small part because gray whales have been virtually ignored by scientists.

That's not a condition unique to grays. What we know about whales—even orcas, which we've been capturing and keeping for entertainment, and ostensibly for study, since 1961—is absolutely dwarfed by the really basic things we don't know. Why do whales sing? How do they find their prey? How and why did they get so big? We've got some great theories, but no concrete answers.

Still, when Torres traded New Zealand for the Pacific Northwest, she was amazed to find that research on the grays was so thin.

"It really shocked me when I got to Oregon," Torres said as Pancake surfaced a few yards from our boat, rolled to her side, and lifted a flipper as though asking for a high-five. "As you can see, they're really accessible, but not very many people had seen that as an opportunity for us to learn more about them."

And that's too bad, Torres said, because what we learn about grays can help us understand other whales that are much harder to find. Even the resident blues of New Zealand present a much bigger challenge for researchers, she said.

Torres thinks the scientific disinterest stems from the fact that grays have rebounded quite well from the point of near extinction in the 1950s; *Eschrichtius robustus* was removed from the Endangered Species List in 1994. That certainly doesn't make them uninteresting, she said, but it does take away the sense that the research needs to be done now, before it's too late.

But given how hard it is to study whales in general, Torres said, the grays—who live their lives a lot closer to the shore than most of their cousins—can offer us a vital window into the risks faced by cetaceans as a whole, including those who are in much greater danger of extinction.

To that end, Torres had invited me to join her and Chandler as they spent a day observing Oregon's "resident" grays, making identifications, taking measurements by way of drone video footage, and . . .

. . . scooping up poop.

That last part was going to be my job.

"So, um . . . how do I do this, exactly?" I'd asked her as we got into position behind Pancake and waited.

"We'll only have a short time to get it, about thirty seconds before it dissipates, but thankfully, Todd is really good at steering the boat right into it," Torres said, pulling a small net from a bucket. "Just reach out into the water, swirl the net around, and try to get as much of it as you can, because we might only get one pass at it."

I stared down at the net. The handle was only about 18 inches long—and I'm a tiptoes-to-reach-the-top-shelf kind of guy. It occurred to me that in order to scoop up the whale poop, I'd have to lean over the side of the boat and bend my torso right into it.

"Is it . . . gross?" I asked.

"It's not too bad. Blue whale poop is actually a lot worse. It's runnier," she said. "But what we get out of it makes it totally worth it."

From a research perspective, whale poop is gold. From it, we can get to know what whales are eating, we can figure out if they are pregnant or nursing, we can do genetic sampling, and we can monitor hormone levels. The fecal samples Torres and her team have been collecting for years are checked for a variety of hormones including cortisol, which is the key stress-response regulator in humans and whales alike.

The reason we know stress hormones work the same way in humans as they do in whales is a very sad one. Back in September of 2001, researchers from the New England Aquarium were studying right whales in the Bay of Fundy, between Nova Scotia and New Brunswick off the easternmost point of Maine, when, as Alan Jackson later sang, "the world stopped turning."[45]

Now, the world didn't actually stop turning.[46] What did stop spinning around, though, were the propellers of commercial ships in the western Atlantic Ocean as the United States and Canada shut down shipping traffic to thwart potential follow-on attacks. When that happened, the researchers—who, like Torres, had been collecting whale feces—saw an immediate drop in the whales' glucocorticoids, the class of steroid hormones that includes cortisol,[47] demonstrating the tremendous impact maritime traffic has on whale health. That finding alone set the stage for a fascinating set of questions about how commercial and industrial activity impacts the stress levels of all animals, including humans.

Now, Torres is working to build upon those findings. Her colleague, Joe Haxel, has stationed hydrophones at various places where Oregon's resident whales congregate. Since shipping noise can vary greatly from day to day, the constant monitoring and steadfast poop-collection efforts will enable the research team to see what happens to the whales' stress levels as commercial activity rises and falls.

The key, though, is getting the poop. We followed Pancake for about forty-five minutes without any luck in this regard, and then moved onto another whale, and another, and another. As the day went on, I went from dreading my assigned job to fretting I wouldn't get to do it, for I realized that, gross as it was likely to be, there's no better way to start a story than "So there I was, picking up whale poop."

Alas, the whales didn't cooperate. So here's the best I can do: "So there I was, awash in whale snot."

One of the most important goals of Torres's research is to connect fecal samples to specific whales. To do this on gray whales, the research team needs to get photos, which can be used to connect the animals' monochrome patches, scratches, scars, and barnacles to a database of previously identified animals, like our friend Pancake. To be as certain as possible that they're not observing similarly patterned animals, Torres's team always tries to get photos from both sides—another job I got tasked with that day—and that often means they have to put the boat downwind of the whales.

Torres and Chandler knew what that meant. I didn't. So when Pancake surfaced, shoreside and just yards from our boat, and sent a misty cloud into the morning air, and I looked over to see both of them dipping their heads into their arms, I just figured I was witnessing a strange inside joke.

And then it hit me: a slimy wet blanket of air with the stench of rotten fish and bile.

Just about everyone who studies whales has had this experience. And whale blow, also known as exhaled breath condensate, or EBC, can hang in the air for quite a bit—so long that sometimes you can even smell an animal before you see them.

This shared experience inspired Iain Kerr, the chief executive officer of Ocean Alliance, a whale research and education organization based in Massachusetts, to think about other ways of collecting biological samples from whales.

The reason why EBC smells so bad, Kerr realized, is because it's chock full of chemical and organic material. With each mighty blow, a whale sends a huge plume of carbon dioxide into the air, carrying with it phlegm, microbes, and even little loosened pieces of whale flesh.

What, he wondered, might we learn if we could capture it?

A few scientists had tried to do this before—driving their boats within feet of their research subjects and holding long sticks, sponges affixed to the end, over their blowholes. It was incredibly hard. It was dangerous. And it was likely stressing out the whales.

Like many research organizations, Ocean Alliance had long ago figured out the benefit of drones for video observation. And as he watched the increasing ease, accuracy, and speed at which drones could be flown, Kerr had an idea. He attached specimen-collection sponges to the miniature aircraft and began flying them over the whales his organization studies—right into the EBC.

Thus, SnotBot was born.

It didn't take long for SnotBot to start demonstrating its merits. In its first major expedition, the drone captured EBC samples from blue whales—and a University of Alaska marine biologist named Kendall Mashburn was quickly able to identify both cortisol and progestogens, giving scientists a new way to test whales for stress and reproductive status. Torres later told me SnotBot was giving her reason to wonder if she would soon be able to spend a little less time following Oregon's resident grays in wait of all that magical poop.

SnotBot is a great example of how scientists are applying relatively cheap technology to attack some really big challenges. It's also proof that people who aren't scientists are eager to support science—especially when a superlative organism is involved. When Kerr wanted to fund his SnotBot project, he turned to Kickstarter, and quickly met his fundraising goal of $225,000 with donations from more than 1,700 individuals.

Talk to any scientist for long, and eventually the subject will turn to money. Research funding cut during the last global recession didn't rebound in many areas when the economy took a turn for the better. After surveying 11,000 researchers in 2014, the *Chronicle of Higher Education* reported that nearly half had abandoned research they considered to be "central" to their mission because of lost funding.[48] The problem, the report's writers concluded, is that basic research "often seems to have no immediate payoff."

That perception isn't wrong. Foundation-laying research, of the sort we're just now coming around to doing on extreme creatures like elephants, giraffes, and whales, won't generally give us novel cures for diseases or mind-blowing revelations about our universe. But none of that other stuff is ever going to happen unless we lay that foundation.

A lot of tiny donations aren't likely to level the funding field. But the success of projects like SnotBot can give us a clue about how to get people excited about science. That, in turn, can help us better understand how to galvanize public interest and support in a bid to compel legislators and policymakers to act.

And, as Torres noted, nothing excites people quite like megafauna.

We don't have to stop at megafauna, though. For the world's biggest organisms aren't fauna at all.

HOW THE WORLD'S TALLEST TREES ARE FIGHTING GLOBAL WARMING

An earthy face appeared over the ledge of a plywood platform suspended 170 feet above the forest floor. "You just climbed my wife," the scruffy-bearded man said. "Did you know that?"

There were a million questions I could have asked. Something about the consummation process of this intra-eukaryotic marriage, perhaps? But after a strenuous climb to what was just over the halfway point of this colossal conifer, very little came to mind.

"So you're the one they call the Lorax?" I asked.

"Yes," the man replied. "I speak for the trees."

The tree-sitters I met in Fall Creek, in central Oregon, were a fascinating lot. There was the Lorax, who had indeed exchanged vows with a tree named "Grandma" on the night before my arrival, and swore he heard his bride say "I do." There was Skye, who was dashing naked through the forest when I first rolled into a camp the eco-protesters called "Red Cloud Thunder." There was Sage, a nineteen-year-old East Coaster who hitchhiked across the country to join the protest when he learned the old-growth forests in the Pacific Northwest were at risk. They called themselves Ewoks, and starting in the spring of 1998, when the US Forest Service sold the logging rights for these woods, they had spent their days and nights in the trees in an effort to prevent loggers from cutting down some of the largest life-forms on our planet.

Sometimes, police and prosecutors would later allege, the protesters gathered at Fall Creek plotted attacks against what they considered the "corporate state" on behalf of the Earth Liberation Front. A year after I first visited their camp, two former Fall Creek tree-sitters, Craig "Critter" Marshall and Jeffrey "Free" Luers, were arrested for firebombing a Chevrolet dealership in the nearby city of Eugene. "If one in 10 people care about the planet," Marshall later told the *New York Times Magazine*, "that one person has to do 10 times as much as those other nine."[49] He was right about that—even if he was so incredibly wrong about how people should go about inciting change. The tree-sitters were already seen by most people as kooky. In the wake of the firebombing, they started to look like something else: terrorists.

And, as it turns out, we don't need violence to save the world's biggest trees. They're making a perfectly good case for themselves.

In fact, they might yet save us.

There are roughly 7.5 billion people on our planet. And there are, according to one assessment from Yale University, just over 3 trillion trees.[50] Forget for a moment the notion of biomass; in numbers alone we are vastly outnumbered.

OK, now *don't* forget the notion of biomass. Look out the window at the nearest tree. On the off chance it happens to be smaller than you are, it won't likely be for long. And it will likely outlive you, too. An apple tree can live for a hundred years and more. An elm can live to be 200. An oak can live for 300 years. And each year these trees grow bigger, and bigger, and bigger still.

The tallest members of the kingdom Plantae only account for a small fraction of the total number of trees on our planet, but we have begun in recent years to recognize just how big of a deal they are in regulating atmospheric carbon.

It stands to reason that old-growth redwood forests should be good at sequestering carbon. They can grow to heights greater than 300 feet—the tallest, known as Hyperion, in Redwood National Forest, has been measured at nearly 381 feet—and they can live for more than 3,000 years. In every second of their lives, they are taking in carbon dioxide from the air and locking it away in their heartwood where it will remain even hundreds of years after the trees fall.

We've been aware of the impact of carbon on global warming since the 1960s, and it has been decades since a widespread scientific consensus emerged about the role of heat-trapping greenhouse gases in causing climate change. And yet no one attempted to measure just how much carbon these giants are taking in until 2009.

Doing this sort of research isn't easy. To make it happen, a team of scientists from Humboldt State University and the University of Washington examined eleven redwood forests in California, meticulously measuring every tree and shrub—not just the towering redwoods, but everything below them, too. They ran samples of leaves, bark, and heartwood through an elemental analyzer, revealing how much of each sample was made up of carbon.[51] They then used computer models to estimate the number of needles on each tree. And then, just to make sure they were right, they actually counted the needles on some of the trees. The effort took seven years.

The results were stunning. No known forest in the world is capable of storing so much carbon as the redwood forests. Not the other conifer woods of the Pacific Coast. Not the old-growth eucalyptus stands of Australia. Not even the tropical rainforests that rightfully get so much attention in the popular conservation movement. If our planet's forests were banks and carbon were cash, the giant redwoods would be the US Federal Reserve.

And here's the kicker: It appears the carbon-rich atmosphere humans have given this planet is actually good for redwood growth. The researchers found that as carbon dioxide levels have risen in recent decades, so too have the redwoods.

That doesn't mean we can just let these trees do their thing and everything will be fine, especially since we've already destroyed about 95 percent of old-growth redwoods.[52] There are no simple answers to vastly complex problems like climate change. But we shouldn't succumb to the notion there is nothing we can do, either, for it's actually not hard to make a difference—and you don't have to live in a tree or firebomb a Chevy dealer to do it.

Want to do something significant to offset your carbon footprint? Plant a redwood, or contribute to conservation organizations like the Redwood Forest Foundation or the Save the Redwoods League. If you do, you'll be helping to save the biggest carbon-sequesterers in the world.

But not the biggest plants. There's actually something even bigger. Much bigger.

HOW THE WORLD'S LARGEST PLANT WAS DISCOVERED, THEN FORGOTTEN, THEN DISCOVERED AGAIN

Burton Barnes didn't want to make a big discovery.

He wanted to make a small one. And then another. And another after that. Over time, Barnes figured, all of those little discoveries might add up to something. Or they wouldn't; he was fine with that, too, for he'd spent a very happy lifetime uncovering little secrets about the world without anyone else paying much mind.

Barnes spent his childhood in the rugged pine forests shadowing the shores of Pokegama Lake in northern Minnesota, where his father worked as an art teacher at Camp Mishawaka, and in the thick beech-maple woodlands to the east of his boyhood home in Charleston, Illinois. He went camping and spent his days collecting, pressing leaves and flowers, and filling meticulously detailed notebooks with descriptions of the flora he found.

Those books might have been the full extent of Barnes's scientific adventures had it not been for his love of music, which led the avid trombonist to the University of Michigan Marching Band in the early 1950s, and then to a rather unusual German fellowship, specifically intended for a forestry student who was also a consummate musician, to study at the University of Göttingen, just miles away from the dense spruce woods and peat-moss-covered bogs of the Harz Mountains.

Barnes returned to the United States in 1959, on the centennial of Charles Darwin's *Origin of Species* and at the dawn of the modern era of human genetics. It was a time when Canada had launched a widespread eradication program aimed at replacing the quaking aspen—*Populus tremuloides*, the most widely distributed tree in North America—with much more marketable conifers. And having come to appreciate during his time in Göttingen the elegance and importance of forests of all kinds, even those made up of what was widely considered a "weed tree," Barnes set out to better understand the aspen, one small discovery at a time.

For a species so incredibly widespread—you could hike east to west across North America and barely leave the sight or shade of a quakie—there had been remarkably little study dedicated to the aspen. No one even knew how big they could get.

In fairness, that is a tough question to answer. Aspen are clonal. They spread below ground, crawling just beneath the surface of the earth through a unified root system, stretching out for water and reaching up, occasionally, for sunshine, through their stems, which most folks would call their trunks. If you see two aspen stems close together—or three or four or twenty, for that matter—you are likely looking at a unified, single genetic colony, one that might be greater in mass under the ground than above it.

Barnes had learned to classify and map forests during his time in Germany. And he'd developed a rather keen eye for subtle clues that would allow an ecologist, such as himself, to map the independent clones within a forest. Barnes traveled across North America, studying the colors and patterns on trees' leaves and bark and comparing his notes to aerial photos, to get a better idea of how big an aspen colony could get.

Near Coot's Slough, off the southern tip of Fish Lake in Central Utah, at an elevation of 9,000 feet, Barnes found a likely answer. He drew a perimeter around a 107-acre colony and, having done so, planted a superlative flag upon the Earth, for if the clone he found there was truly that big, it would be the largest organism ever discovered. And not by a little.

An entire herd of elephants could live under its shade. From mouth to tail, a blue whale is just barely longer than one fully grown aspen stem—and this clone has 47,000 stems. If all of its 380 feet were to fall to the ground, Hyperion, the world's tallest-known redwood, would not even reach across the clone's width. And at an estimated weight of 2.7 million pounds, what is thought to be world's heaviest-known sequoia, General Sherman, is likely just a fifth as heavy as the Fish Lake aspen clone.

The clone Barnes discovered is not just big; it's mobile. Over the course of time, an aspen colony can migrate from one place to another as it seeks better soil and exposure to the sky. And sometimes, in the midst of this slow subterranean crawl, a part of the clone can become separated from the master colony by a landslide, fire, or human intrusion. Like conjoined twins split by a surgeon's knife, the parts remain genetically identical to the whole. So it's possible something similar happened to this clone—a two-lane road runs right through its center—and what Barnes found is not one organism anymore, but two. If so, the separated twins would still likely be the first- and second-largest known plants in the world.[53]

Barnes could have put his name on the Fish Lake clone, as discoverers often do. Instead, in 1976, he buried a report about his superlative discovery amid other aspen data in an obscure Canadian scientific journal. He would later say that the discovery was little more than an outlier. All he'd done, he told me in a brief correspondence in 2013, was identify "an atypical example" of one of the world's most common trees.

When Barnes died the following year, his obituary didn't even mention the Fish Lake aspen. And because Barnes' discovery wasn't particularly well known, even among forest scientists, the clone went virtually unstudied and tragically unprotected for many years.

"There were campgrounds and cabins and firewood-cutting areas inside the clone," said Michael Grant, a professor of ecology and evolutionary biology at the University of Colorado in Boulder, describing his first visit to the Fish Lake clone more than a decade after Barnes identified it as the largest known organism in the world. "It wasn't marked. No one was doing anything to highlight it. There was nothing to signify that it was an important natural wonder." The clone's stark stems, he said, had become a gallery of arborglyphs. The carvings, concentrated around campsites, were mostly names and initials. But there were also peace signs, scriptural citations, happy faces, and crude pornographic sketches. Each carving invited insects and disease.

Grant came to see the clone's broad anonymity as a threat to its safety. And in 1992 he came upon a way to change that. Writing in the journal *Nature* that year, a team of Canadian and American researchers had bragged they had found the world's largest singularly genetic organism—a 38-acre fungus growing on the roots of trees in Michigan's Upper Peninsula. Not to be outdone, the US Forest Service and the Washington State Department of Natural Resources countered with their own discovery—a 1,500-acre fungus south of Mount Adams, weighing in at an estimated 825,000 pounds. But bettering them all in *Discover Magazine* the following year, Grant laid out the case for Utah's enormous aspen clone. And reasoning, as many conservationists have, that humans have a harder time destroying things that have been anthropomorphized, Grant gave the clone a name.

He and his colleagues called it "Pando"—Latin for "I spread."

"It was simple. It was easy to say," Grant said. "It had nice phonemes. It fit the situation reasonably well. I'm sure there were a lot of other things that would work, but that's what we went with."

The name stuck, even as Pando's claim to modest fame came under increasing scientific skepticism, and suffered under the general scientific malaise that all too often surrounds biological outliers.

Then, from 2000 to 2006, a widespread drought claimed up to a fifth of the aspen in some areas of the American West, resulting in an unprecedented collapse of the biodiversity that was supported by these colonies. Suddenly, the largest known aspen clone was scientifically interesting—specifically *because* of its size. "If you want to know under which conditions aspen thrive," conservation geneticist and molecular ecologist Karen Mock told me in 2013, "it's certainly worth examining the one that appears to have thrived the most."

But Mock wasn't convinced that Pando was, in fact, the world's largest known aspen clone. In fact, she told me, she strongly suspected Barnes' 107-acre estimate was wrong. To know for sure, she needed to take a look at the purported clone's DNA.

Before she could get to that task, she had to collect a lot of samples. Using small pieces of barbed wire, ice fishing rods, and her bicycle-helmeted children as test casters, Mock and her family practiced fishing for leaves in her backyard. It wasn't easy. "Sometimes, the leaves are way bloody up there," she said. Ultimately, they ditched the rods and barbed wire in favor of slingshots. Then they dried out the leaves they collected using kitty litter and crushed them into a powder to analyze the DNA.

Populus tremuloides has one of the shortest genomes among trees, with just 550 million base pairs. That's about forty times fewer than a common pine. "But that's still a lot of DNA," Mock said.

When the data-crunching was done, Mock mapped the results. And instead of "taking down Pando as a construct," as Mock expected would be the result of her work on the Fish Lake clone, the genetic tests showed Barnes's decades-old map, based on nothing more than aerial photography and his own eye for detail, was almost perfectly aligned to a map based on the tree's genes.

"It was almost like a tracing," Mock marveled.

By paying attention to small details, Barnes had indeed discovered the world's largest known plant.

What secrets might a behemoth like this hold? What scientific insights might it offer us? We don't yet know. Like so many other superlative life-forms, this enormous aspen has gone largely unexamined. Even after Mock confirmed its gigantic size, only a handful of studies about it have been published in peer-reviewed journals.

The mysteries that remain are huge.

Chapter II

ALL THE SMALL THINGS

Why Little Organisms Have Such a Big Impact on Our World

Even knowing what it was, and generally where it was in relation to the town of Rifle, Colorado, I blew right by it on my first pass. And my second, too.

I finally found what I was searching for down a steep driveway, tucked behind a racetrack for remote control cars, and not-so-well guarded by a cattle gate and wire fence. I ducked between the wires. It felt like I was breaking into a temple. It felt holy.

The Rifle Integrated Field Research Challenge site was, alas, no more exceptional to look at than any other abandoned patch of land along the Colorado River. It was June, and the grass was tall, brown, and brittle. The sticker bushes left scrapes on my calves, and I was thankful that I'd swapped my flip-flops for boots, but regretted not also trading my shorts for jeans.

Amidst the scrub grass there were some white pipes, about six inches in diameter, sticking up a few inches from the ground—wells for pulling water from deep below the surface. There was also a small solar weather station, a portable single-wide trailer like the ones construction crews use for onsite offices, and a tool shed. That was about it.

The little piece of land seemed like it might be a nice place to go fishing. Or maybe to have a picnic, if you cleared out a blanket-sized patch of the weeds—and if you could forget that this plot was once covered by nearly 5 million dry tons of radioactive milling waste.

The waste—fine sand known as tailings—was in large part left over from the Atomic Energy Commission's wild rush for nuclear fuel in the 1940s and '50s. It had sat there for decades, soaking into the ground by the river and leeching its contaminants into the groundwater.

Back in the mid-1990s, the US government began a four-year, $120 million process to move the tailings, but the damage was largely done. The vanadium, selenium, and uranium left behind in the soil and water were supposed to dissipate under "monitored natural attenuation." Two decades later, the pollution endured.

And so the government went looking for ideas. Even crazy ones.

A motley group of geologists, chemists, and biologists from around the country recently had been crafting papers with titles like, "Detection and characterization of a dehalogenating microorganism by terminal restriction fragment length polymorphism fingerprinting of 16S rRNA in a sulfidogenic, 2-bromophenol-utilizing enrichment." In English: Identifying microbes that could be compelled to breathe in one thing and breathe out another.

Maybe, it was suggested, they could find a bacterium that breathes in uranium.

That was indeed a crazy idea. Bacteria consume a lot of different things, but radioactive waste wasn't on the list. Not that we knew of, anyway. Up until very recently in the planet's history there just wasn't very much of it lying around. But the researchers knew the uranium had been seeping into the soil at the Rifle site for a long time. They knew there were a lot of bacteria down there, too. There's bacteria everywhere, after all. Maybe one had developed a habit of huffing uranium. And maybe, using the same techniques they'd been using to get bacteria to respire a whole bunch of other stuff, they could help feed its habit, thus helping to clear the remaining uranium from the Rifle riverside.

This was the idea that brought a Justice League of scientists to the middle of Garfield County, on the banks of the Colorado River. This was how the Rifle team got its start.

The Rifle researchers certainly were not the first people to suggest bacteria could play a positive role in the world. That's fairly common knowledge now, though it took us a really long time to get there.

From not long after an amateur scientist named Antonie van Leeuwenhoek[1] first described the world of the "small little animals" he saw in the homemade microscope he created in 1671, we've treated bacteria as a terrible scourge. That's in no small part because bacteria can, in fact, be a terrible scourge. Tuberculosis. Lyme disease. Cholera. Typhoid fever. Take your pick; the world of bacterial diseases is a whole lot of rotten.

Yet lately, we've been learning that there are good bacteria and bad bacteria. And in our efforts to vanquish the bad ones from our bodies, Earth Microbiome Project co-directors Jack Gilbert and Rob Knight wrote in 2017, "we have inadvertently unleashed a Pandora's box of modern plagues—the array of slow-killing, miserable, chronic health problems that have become prevalent across the modern world." Asthma, allergies, and rheumatoid arthritis. Celiac disease, irritable bowel syndrome, and multiple sclerosis. Even some mental illnesses.[2]

A rapidly emerging field of science is demonstrating how bacteria can help us address all of those illnesses, and more. It is quickly becoming clear that bacterial transfer—putting those good bacteria to work against the bad ones—could be the therapy of the future. It might not be long before we're hearing doctors say, "Take two bifidobacteria and call me in the morning."

Meanwhile, we're learning the presence of bacteria in the world outside of our bodies can be very healthy for us, too—an understanding we owe in no small part to the Cold War, which separated West and East Germany into rival states of "haves" and "have nots" for more than forty years. When the Berlin Wall finally fell in 1989, health researchers like Erika von Mutius recognized an opportunity to study the impact of germs on children raised in the "clean" West and the "dirty" East. von Mutius expected children from East Germany would have more allergies and suffer from asthma at greater rates. The opposite was true. That finding was essential to the development of new field of inquiry, dubbed "the hygiene hypothesis," suggesting human immune systems derive tremendous benefit from being exposed to plenty of microbes.

So yes, some germs are bad. And yes, some germs are good.

The thing about germs, though, is that it's really hard to kill off the bad ones without killing the good ones, too. Penicillin and Lysol aren't particularly discriminating. And perhaps a bigger problem, in the end, is that not every germ is good or bad. Some are both, and not just for our health.

Take the class of tiny critters called betaproteobacteria—horrible germs that can cause meningitis, pertussis, and gonorrhea. A bacterium from this same class was among the hundreds drawn from the wells at the Rifle site—and the team soon realized this microorganism had somehow evolved to breathe in radioactive uranium, stealing electrons in a process called "reduction" that leaves the output product less radioactive.

The germ class that gives us one of the most prolific sexually transmitted diseases in the world, the researchers hypothesized, might also help us rid the Earth of some of its radioactive waste. If that's not the backstory for some truly gnarly dystopian fiction, I don't know what is.

That was, I admit, the thought crossing my mind when my phone rang and the screen registered a number I'd added to my address book just a few days before coming out to Rifle: "Kenneth WILLIAMS (DR.) RIFLE SITE." The co-author of the uranium-breathing bacteria paper—and manager of the research site where I was standing—was calling.

I looked up at the roof of the portable building. A camera was looking back down at me.

"Doctor Williams," I answered with the same eager inflection children use to address their parents when they get caught with a hand in a cookie jar. "How are you, sir?"

The connection wasn't great. I only caught every few words, but I could make out one question distinctly: "Where are you right now?"

I figured he was being coy, but when I fessed up, he laughed. No one was monitoring the camera—and it seemed to amuse him that I believed he had enough funding to have someone watch surveillance footage twenty-four hours a day. Apparently discovering a novel way to potentially rid the world of radioactive waste is not a ticket to a fat research bankroll.

He was, in fact, just returning my call from earlier in the week. I'd told him I was hoping to meet up while I was in the area, and he wanted to get together, too. There was a lot to talk about, after all. Because as amazing as the uranium-breathing germ is, it's probably only the second most important discovery that came out of Rifle.

"You know," I told Williams on the phone, "this place could really use a bronze plaque. Something like, 'Here we changed life science—forever.'"

"I was thinking that same thing," he said.

HOW MICROBES HELPED SCIENTISTS DRAW A NEW TREE OF LIFE

Rifle, Colorado, is where our entire world changed—the place where the tree of life was uprooted, cut up, and put back together like a Salvador Dalí version of the original.

To understand how this happened, we've got to go back to 1837, when Charles Darwin was re-establishing himself in Cambridge after his nearly five-year journey of discovery on the HMS *Beagle*. Contemplating his emerging theory of "transmutation," what we've come to know as "evolution," he sketched out a rudimentary tree of life. In it, he illustrated his belief that the diversity of creatures on the planet, and before them a "long succession of extinct species," leads back to a common ancestor. "From so simple a beginning," he famously wrote at the end of *On the Origin of Species*, "endless forms most beautiful and most wonderful have been, and are being, evolved."

Typifying the humility I've come to appreciate and expect from most (though by no stretch of the imagination all) scientists, next to the tree Darwin had scribbled, "I think," leaving room for his hypotheses to evolve, too. Sure enough, we've been adding to the tree ever since, which is what microbiologist Carl Woese was doing in 1977 when he proposed a classification system that separated all living things into the three-domain structure most of us become familiar with right around middle school—Archaea, Bacteria, and Eukarya.

Genomic research has since offered us new ways to draw evolutionary links between animals, like the African elephant and the rock hyrax, which don't look closely related but are. And as we've learned these things, we've redrawn the tree—usually species by species, and very slowly.

Then Jillian Banfield came along. She'd spent her career studying and sequencing microbes, taking particular interest in ones from some of the most inhospitable places in the world. Among her favorite research sites is a series of caverns in Northern California where temperatures can reach nearly 120 degrees Fahrenheit and the natural groundwater is the most acidic known on Earth.

In Rifle's uranium-laced groundwater, she'd found another hellish playground.

Researchers often filter water to get larger contaminants out, allowing for an easier search for the things they're actually looking for. On what seems to have been a whim, though, Banfield just decided to keep filtering, using successively smaller filters, well past the level usually used to sterilize water. Most bacterial researchers wouldn't have done such a thing, because the assumption was you'd eventually filter out all the bacteria, which would defeat the purpose. When she and the Rifle team scraped off the screens and sequenced what was there, though, they discovered 789 sets of genetic code relating to organisms as small and perhaps even smaller than anything previously identified—a whole new world of the tiniest creatures on Earth.

Figuring out where all those organisms belonged on the tree of life proved challenging. There just wasn't a good shelf for them. And these are really big, accommodating shelves—the sorts on which you could place an elephant and a hyrax and say, "Yeah, that totally goes."

Banfield's team found homes for the new organisms in seven existing phyla. For the rest, they had to create 28 entirely new phyla. The DNA was just that different. "This wasn't like discovering a new species of mammal," microbiologist Laura Hug later reflected. "It was like discovering that mammals existed at all, and that they're all around us and we didn't know it."[3]

The Rifle discovery nearly doubled the number of branches in the bacterial domain and further dwarfed the archaea and eukaryotes on the tree of life. The branch we're perched upon—the one Darwin originally sketched out in his notebook—doesn't look like much more than a twig, now. And the discoveries have only just begun, which means we're soon to be even further dwarfed by what we once believed to be our dominion.[4]

And here's the thing: It almost didn't matter that the world-shattering discovery of these bacteria happened to occur at the Rifle site. There are many extreme environments out there that have yet to be sampled for microbial life. It was the simple act of filtering, along with some novel ways of searching and sequencing the DNA, that gave us an entirely new way of looking at the vast diversity of life. A discovery like that could have happened anywhere.

That doesn't make Rifle any less awe-inspiring. Not to me, at least.

I was twenty when I first visited Jerusalem, and in that trip I made my way to the Church of the Holy Sepulchre, where many Christians believe Jesus was killed, buried, and resurrected. There are far more elaborate shrines and sanctuaries out there. There is Barcelona's Sagrada Família, Moscow's Saint Basil's Cathedral, and London's Westminster Abbey, marvels of architecture every one. Christ, it is told in all of those places, was rabble-rousing all over Judea, Samaria, and Galilee; he could have been crucified anywhere. But a crumbling, gray-stone church in the Christian quarter of Old Jerusalem is, if you believe the stories, where everything changed.

Unassuming though this scrubby little patch of land on the Colorado River is, Rifle is the place where life science took a fundamental leap forward, and nothing will ever be the same again.

WHY MICROBIOLOGISTS ARE LIKE CRIMINAL PROSECUTORS

There's a chance we're wrong about blue whales. It's a slim chance. A chance on the order of small fractions of a percent. But it's a chance nonetheless.

Maybe *B. musculus* isn't the largest animal ever to live on our planet. Maybe there was something bigger. The known fossil record doesn't give us any reason to believe this is probable but, in the field of paleontology, wonders truly never cease.

So yes, it's possible.

In the other direction of things, though, the situation is completely different. Sure, there's a chance we've already discovered the smallest organism ever to have lived on our planet, but it's a chance on the order of small fractions of small fractions of small fractions of a percent.

So far, we've identified about 6 million different microorganisms. Some scientists, though, believe there could be a trillion different species of microbes out there.[5] To put this into context, imagine every individual bird in the world was its own species. OK, got that in your head? Now double it, and you're almost there.

And that only comprises the microorganisms likely living on Earth with us right now. Jay Lennon and Ken Locey, the Indiana University research duo that published the trillion-species estimate, told me in

2017 the question of how many species have inhabited the planet at some point in the past 3.5 billion years (give or take a few hundred million) is one they'd love to get to. At that point, along with geologist Kevin Webster, they were exploring the possibility of using the Sepkoski Curve, which describes the accumulation and extinction of eukaryotic taxa over the past 600 million years, to get a better sense of what has happened to other organisms over those eons. "If we can understand the statistical properties of Sepkoski Curves, then we can potentially—and cautiously—apply that to microbes and then back-track it a few billion years," Locey told me.

But even among the microbes that have been discovered, a lot haven't yet been seen.

How do you discover something you can't see? The same way prosecutors file charges when they don't yet have a human suspect in a sexual assault case—by charging the person's DNA. In the fall of 2015 in Seattle, for instance, prosecutors filed felony charges against "07-3116 Peri-SF," a DNA sequence that had been sitting in an FBI database for eight years, in an unconventional approach to beating the ten-year statute of limitations for prosecuting rape in Washington.[6] So far, such tactics have mostly been used in hopes of matching future DNA samples to "cold cases," but, in the not-too-distant future, it could be common for investigators to develop criminal profiles—and even detailed composite pictures of individual suspects—based on DNA left at a crime scene. A scientist named Mark Shriver and his team from Penn State University came up with a process by which thousands of DNA markers for facial attributes—such as skin, hair and eye color, and nose, lip, and eyebrow shape—can be used to build a fairly convincing composite sketch of a person, based only on pieces of their genetic code.[7]

In Rifle, Colorado, Banfield and her team did essentially the same thing.

Imagine, if you're not squeamish, what a human body would look like after being put into a food processor the size of a minivan. The microorganisms Banfield's team collected were in that sort of shape—the DNA was in bits and pieces. Like paleontologists piecing together small bits of bones to assemble bigger pieces of bones, then putting those big pieces together to get a better idea of what a creature might have looked like, the Rifle researchers identified overlapping segments of DNA to create longer sequences.

However, in all but a few cases, they didn't have the complete genomes. In most situations, they had about 90 percent of each microorganism's code—enough to know that each was, indeed, a unique species not previously identified, but not enough to create a perfect composite picture.[8]

The team did get a few amazing photos of some of the microbes they found. Using a portable cryo-plunger, which looks like a cross between a blender and a table-top wine opener and can be used to flash-freeze specimens at –272 degrees Celsius, the researchers were able to stabilize some of the bacteria they pulled from the Rifle wells long enough to get them to a lab. There, they used a souped-up electron microscope, capable of taking images with a pixel size of about one-fifth of a billionth of a meter, to take pictures. Even with the high-powered equipment they used to get pictures of the microbes, the images are fuzzy—like the picture on an old black-and-white television with a tiny antenna. Banfield later told me it's not possible to say exactly which organisms they captured in the photos—that's going to take a longer and very elaborate process of connecting phenotypical traits to genetic sequences. Nonetheless, when all was said and done, the team had several images of single-celled organisms whose size could then be measured.[9]

Up to that point, the smallest known organism on Earth was a toss-up. Guinness had long held that an archaeon called *Nanoachaeum equitans*, an "extremophile" living in salty, acidic, and very hot marine environments, was the "smallest entity universally recognized to be a living organism."[10] The cloud-connected crowd of editors on Wikipedia, meanwhile, gave the title to *Mycoplasma genitalium*, a parasitic bacterium that hangs out in the human gut, bladder, and respiratory tracts. Both organisms grow to be somewhere in the range of 300 nanometers across. To the extent scientists got involved, they seemed to have sided with the mycoplasma,[11] although they mostly stayed out of it altogether. There were just too many unknowns to make a hard claim.

Now, thanks to what happened in Rifle, the picture is even murkier. The images obtained there showed bacterial organisms roughly 250 nanometers across. (By way of scale, the period at the end of this sentence is about 1 million nanometers across.) And these were just a small selection of images of random microbes from the Rifle wells. There will be more in coming years, as an entire nanoscopic world, nearly unimaginable until now, comes into better focus.

They won't likely be coming from Rifle, though. When I met up with Ken Williams in his home in Crested Butte, Colorado, a few weeks after my illicit visit to Rifle, he told me the "holy" facility I'd trespassed upon was being deconstructed. "As we speak right now," he said, "they're tearing down the columns of that temple."

We may never draw another sample from Rifle. But among the billions upon billions of yet-undiscovered microorganisms throughout the world, it's almost inconceivable that we won't eventually find something even smaller than what was found there. Then, something smaller than that. Then something smaller than that.

And that will bode very well for all of us. Because when it comes to potential for unlocking the secrets of life—which in turn may help us sustain life—the 2,000-year-old words of the Roman naturalist Pliny the Elder couldn't be more apropos.

"In the whole of nature," he wrote, "there is nothing as great as what is smallest."[12]

HOW THE WORLD'S SMALLEST LIFE-FORMS ARE HELPING US SOLVE A RUBE GOLDBERG PUZZLE

"Size matters not."

I've been repeating those words to my daughter, who is almost always the smallest in her class and on her soccer team, for a very long time. (Also, "Do or do not—there is no try" and "Never tell me the odds," for Star Wars is replete with valuable life wisdom.) Indeed, when it comes to our genes, Yoda was right. Size can be deceiving.

For instance, when an international group of scientists[13] sequenced the genome of the water flea, *Daphnia pulex*, in 2011, the researchers were stunned by what they found. The tiny crustacean had 25 percent more protein-coding genes than human beings, and those genes could be expressed in all sorts of marvelous ways, resulting in a diversity of body shapes and attributes, like thickened sections of shell that serve as protective helmets and even "neck teeth" that make it harder for predators to swallow.[14] One of the most complex genomes in the world, they had discovered, was on a creature that never grows much bigger than 3 millimeters in length, and often doesn't exceed a half-millimeter, even in adulthood.

But for all its complexity, its genome was still quite short. How? As any good computer programmer will tell you, there's a difference between complexity and length, and a really good coder is one who can do more with less. And the water flea is incredibly well coded; despite the large number of genes in its genome, it has just 100 million base pairs. (By way of comparison, humans have 3 billion.)

In sheer length of code the water flea falls into a trend that has been revealing itself in recent years, as genetic exploration has become cheaper, easier, and quicker. More than 5,500 species have had their genome sizes estimated, and there is a 7,000-fold difference between the shortest and longest codes. As more codes hit the market, so to speak, more studies are showing positive correlations between body size and genome length in many animal groups.[15] That's why, if you have a reason to be looking for a short genetic code, you might want to set your sights on the world's smallest organisms.

And biochemists Craig Venter and Dan Gibson needed a short code, for at the time they embarked on their effort to synthetically re-create an existing organism, in the late 1990s, the processes used for building synthetic DNA could only create a few thousand letters of code at a time. Most organisms, even microorganisms, have many millions of base pairs. And plenty of life-forms, as we know, have many billions.

They found a contender for their experiment among the smallest of the small in size. *Mycoplasma genitalium*, the bacteria Wikipedia had long cited as the world's smallest known organism, had just 580,073 base pairs. It also had only 525 genes, the world's fewest among anything we conventionally call life. But the problem with *M. genitalium*, Gibson later told me, is that it grows sluggishly, which would have slowed down their work.

In a closely related species, though, they found the perfect fit. *Mycoplasma mycoides*, a bacterium from the guts of cows, sheep, and goats, has just over 1.2 million base pairs—keeping it well within the range of the smallest known genomes in the world. And it grows 10 times faster than its cousin.

Using *M. mycoides*'s sequence—like kids following the instructions from a LEGO set—the researchers stitched together more than 1,000 individual segments of code to create a chemically synthesized chromosome, which they then inserted into a naturally grown cell stripped of its DNA.[16] The result was a close replication of *M. mycoides*, save for a few "genetic watermarks" the scientific demigods added as proof the microorganism, whose birth was announced in the journal *Science* in 2010, was indeed their own creation.

Among the most fascinating revelations stemming from that journal article was that, like Frankenstein's creature, life begins for these synthetic organisms with a jolt of electricity that delivers the chromosomes into the cellular shell. I cannot help but imagine Venter, who has a well-known flair for the dramatic, extending his arms to the sky and screaming out, "It's alive!"[17]

In 2016, Venter's team took another big leap using the same small organism. Having succeeded in closely replicating *M. mycoides*, they then began stripping away base pairs, and even entire genes, from their synthetic bacterium to see how low they could go. Eventually they cut the original code by more than half.[18] The result was a creature, called JCVI-syn3.0, with just 531,560 base pairs and 473 genes—a life-sustaining genome smaller than any known in the natural world, bringing us as close as we'd yet come to understanding which genetic building blocks are truly fundamental for life.

It will take many more years of research to know if the DNA fundamental for one form of life is fundamental for all forms of life. But as we acquire this knowledge, we'll be in much better position to understand what genes do individually and in concert. And that, in turn, may help us fill some very big holes in our understanding of the genetic world.

Despite decades of research and the billions poured into sequencing the human genome, after all, there are still broad expanses of our genetic code—the vast majority of it, in fact—we simply do not understand. Even respectable scientists long derided this genetic dark matter as "junk." Today we know it's not junk, not by any stretch of the imagination. Much of it, we've learned, is involved in regulation: helping decide which genes are engaged and expressed at any given time for the purpose of surviving in an unpredictable world. But even though we've developed a better respect for this regulatory DNA, we're not much closer to understanding with any sort of specificity what it all does.

That's because, in a lot of ways, our DNA is like one of those zany Rube Goldberg machines people build to complete a simple task in an elaborately complicated and whimsical way. Goldberg's most famous creation (though only depicted in a cartoon and never actually built) was a "self-operating napkin" machine, which involved a pulley, a parrot, a pail, and a rocket. An infinitely more complex machine, featured in a video for the OK Go song "This Too Shall Pass,"[19] employed an old-fashioned typewriter, a piano, an electric guitar, and both real and LEGO versions of a Ford Escort "LeMons" racer[20] to fire four paintball cannons at the band's members.[21]

The role of each isolated part of OK Go's machine, viewed piece by piece, might be tough to imagine—as would the purpose of the finished contraption. But when you view the device being used in sequence from start to finish, the purpose of each piece becomes obvious. Likewise, when genes are added back onto a genome that includes only the bare essentials, their purpose as part of ever more complex organisms may become clear. By starting with what is fundamental—what is absolutely necessary to sustain life—researchers believe we may be able to better understand the purpose of all of that mysterious other DNA.

The principle of starting from the smallest of things to better understand everything bigger is useful for looking at many groups of organisms, from the smallest plants to the smallest insects to the smallest amphibians and the smallest mammals. All of these superlative life-forms can tell us something fundamental about everything else.

If we don't miss them.

WHY OUR MICROBIOMES MIGHT BE AS IMPORTANT AS OUR BRAINS

It was something of an accident that Antonie van Leeuwenhoek even came upon the microbial world in the late 1600s. The Dutchman wasn't a trained scientist. He had been born into a family of basket makers and brewers. As a young man, he had set himself up as a fabric merchant in the town of Delft, where he was a contemporary of the painter Johannes Vermeer. Vermeer is famous for his *Girl with a Pearl Earring*, but his work in paintings like *The Astronomer* and *The Geographer* indicates the reverence for science and scholarship that permeated Delft at the time.[22] It was within this atmosphere that van Leeuwenhoek took up the craft of microscope making, and began an amateur study of the previously invisible world. Perhaps in a different setting, he might have kept what he saw underneath his lenses to himself, for fear of being accused of madness—or, worse still, for fear of actually being mad. Instead, he was emboldened to write to the Royal Society of London for Improving Natural Knowledge to explain what he had seen.

After overcoming some considerable skepticism, the Royal Society's members were initially enchanted by van Leeuwenhoek's discoveries, which included protozoans, bacteria, and spermatozoa. But the novelty of the microscopic world wore thin when no one could decide what, if anything, these tiny life-forms actually did. They could be looked at. That was something. But that was about all they seemed good for.

"It would be almost 200 years before these organisms would garner further serious attention," wrote Paul Falkowski, who researches how organisms transformed the geochemistry of Earth at Rutgers University, in 2015. "Amazingly, while the fundamental discoveries in science in the seventeenth century—gravity, light waves, planetary rotation around stars, and the incredible abstraction of science in mathematics—spurred huge explosions of discoveries in physics and chemistry, fundamental discoveries in biology largely lagged behind and were important only as they related to human health."[23]

In light of what we are now learning about the role of microbes in human health, the irony is rather rich. And it's even richer given that microbes were both the first life-forms on Earth and organisms that, over billions of years oxygenating our planet, created a world where life as we now know it could exist at all.

It's hard to say what we might know if the sorts of revelations taking place in other fields of science during the Age of Enlightenment had extended to the microbial world. But among microbiologists there seems to be a consistent lamentation that we missed out.

"I think that we're still playing catch-up," Rob Knight told me in 2017.

Knight's go-to metaphor for what we lost, by ignoring microbes for so long, starts in a pharaoh's tomb, where ancient Egyptian morticians meticulously preserved their departed leaders' bodies, removing and placing in jars all the organs the pharaoh would need for the afterlife.[24]

Except the brain. They couldn't figure out what that did, so they just stuck a hook up the dead pharaoh's nose, scrambled up all that gray matter, let it drain out, and tossed it away.

We've been doing that same thing, Knight often says, with our microbiome. For a very long time, we knew it was there but didn't really know what it did, so we treated it as though it was unimportant.

Trying to make up for lost time, Knight embarked on an ambitious project: a citizen-funded science initiative to help Americans understand what's in their gut. The effort also resulted in a census of microbiomes across the nation, and later across others. No joke: Knight actually got more than 11,000 people to pay $99 a piece for the honor of sending him a sample of their poop in the mail.[25] Knight also sampled his own stool every day for years—and his exceptionally good-natured partner, Amanda Birmingham, did so as well.[26]

Not all of our bacteria is in our gut, though. A lot of it is in other regions—especially our nether regions. After years of study, Knight knew newborn babies get a lot of their initial helping of healthy bacteria from the birth canal. When he and Birmingham's daughter was born by emergency C-section, they didn't want her to miss out, so they swabbed the baby's whole body—including her ears and mouth—with samples from Birmingham's vagina.

They didn't just do this and keep it to themselves. Knight made a point of detailing the process in a 2014 TED Talk and in a companion book he published the following year.[27] Why? "Because we're practicing for when we tell it to our daughter's prom date," he quipped.

Really, though, sharing his family's potentially embarrassing secrets was about setting a standard. You can't ask thousands of people to trust you with their poop and not trust them with your stories. The reason Knight was successful in getting people to participate in one of the world's largest-ever citizen science experiments isn't just that he's a great scientist—although he is a great scientist. He's also a very genuine communicator. These days, that's just as important.

Already, the American Gut data set has been used to show that adults with nut and pollen allergies tend to have a low diversity of gut microbiota, with lower-than-average populations of Clostridiales (generally a "good guy" bacteria) and higher-than-average Bacteroidales (usually one of the "bad guys").[28] It's shown a potential link between oral bacteria that reduce nitrate, nitrite, and nitric oxide and people who suffer from migraines.[29] And, confirming Knight and Birmingham's postpartum proactivity was a good idea, it has demonstrated that people born by Cesarean section do indeed appear to have a distinctly different microbiome—not just as babies, but long into adulthood.[30]

All of which is to say that when we overlook the smallest things in our world, we do so at our own peril.

HOW THE WORLD'S SMALLEST FLYING ANIMAL BELIES A FUNDAMENTAL CONSERVATIONIST BELIEF

I had figured Burton Barnes' obituary was an outlier. Barnes was a bit of an eccentric, after all, and a tad dismissive of the idea that people outside his particular academic field could appreciate the actual significance of a hundred-acre aspen clone. In his lectures at the University of Michigan, he'd openly derided people who were excited by the idea of a "world's largest organism." So when his biggest discovery didn't show up in the 979-word appreciation of his life that was published in the *Ann Arbor News* after his death in July of 2014, it just seemed like a respectful omission.[31]

Then I realized that the same thing had happened before.

"John W. 'Jack' Beardsley, 74, died Feb. 5 at the Bishop Museum while in town to do the work he loved—studying insects," read the opening line of a 667-word obit in the *Honolulu Star-Bulletin* on February 14, 2001. The story went on to detail many of Beardsley's colossal contributions to the field of entomology, including his more than 650 scientific articles and numerous other published notes on invasive insects in Hawai'i.[32]

But not a word on his smallest discovery.

It was only a year before his death that Beardsley and fellow entomologist John Huber had reported their discovery of a tiny wasp from the Mymaridae family, members of which are commonly known as fairyflies. The species they caught on sticky yellow traps, placed in trees at five to eight feet in height, on Hawai'i, Moloka'i, and O'ahu islands, was as small as 190 micrometers from head to abdomen, about the width of a few strands of human hair.[33]

They called it *Kikiki huna*.

You've heard of the Big Kahuna? This was the little one. The little *K. huna*.[34]

Like Barnes's discovery, the announcement of *K. huna*'s identification was first published in an obscure scientific journal, *Proceedings of the Hawai'ian Entomological Society*. The report—carefully and humbly written as good science should be—didn't overtly make any record-setting claims.

But *Kikiki* was record-setting. The smallest flying insect ever discovered. A completely new genus of fairyfly. And later studies with more collected specimens would reveal *Kikiki* could be even smaller, as tiny as 158 micrometers in length.

When Beardsley and Huber first found *Kikiki*, they thought they'd discovered something endemic to Hawai'i. But, as is the case when it comes to many discoveries, once researchers knew it was possible to look at the world differently, what was previously invisible became clearly abundant. Today the smallest of the fairyflies has been found in Argentina, Australia, Costa Rica, Trinidad, and India.

In India, entomologist Prashanth Mohanraj told me, the discovery of *Kikiki* in one area prompted researchers to begin looking elsewhere. "We were under the impression the *K. huna* is rare in India with an extremely restricted distribution," he said. "But we were pleasantly surprised to see it turn up in our traps from a place where we have collected quite extensively."

Once the scientists started *looking* for *K. huna*, they were finding it in places they'd already searched for lots of other insects.

There are two possible explanations for this.

The first is that they were simply missing it before. Like with the ultra-small bacteria discovered in abundance in Rifle, it wasn't until researchers' how-small-is-small assumptions were set aside that a very big part of the world came into focus. Once that happened, they quickly homed in on some best practices for catching such tiny things—offering other researchers very precise instructions for fairyfly catching that read like a cookbook recipe. "Add ethanol from the sample jar to a sorting dish to a depth of 2–3 mm," Huber and fellow entomologist John Noyes wrote in one paper, specifying that the dish should be "a 9 cm plastic Petri dish with grooves scored at 1 cm intervals on the outside and made visible by drawing black lines in India ink."[35]

The other possibility is that *Kikiki* simply wasn't in some of those places before, and that it had just recently migrated. And that begs a question: Where was *Kikiki* originally from? It's not certain. But a more recently discovered wasp, *Tinkerbella nana*—named for the mischievous fairy and nursemaid dog from J. M. Barrie's Peter Pan stories—might offer a clue.

Tinkerbella, it turns out, looks an awful lot like *Kikiki*, both in terms of body size (Tink is only slightly bigger, on average) and the particular arrangement of veins in both species' wings. They differ, though, in antenna shape, the number of optical units in each eye, and foot joints. As more apparently related species are found, and genomes are examined, we'll have a better idea of where *Kikiki* originally came from. For now Huber believes the Hawai'ian population "was almost certainly accidentally introduced from elsewhere," most likely Central America.

The word "accidentally" is key here. It means *K. huna* likely caught a ride on a human boat or plane. And that, in turn, tells us something really important about the fight to prevent invasive species from overrunning our globalized world: It's too late.

The opinion that nonnative species should be prevented from taking up residence in places where they might change the ecological status quo is widely held among conservationists, and there's plenty of evidence in support of our protectionist stance against species we call "invaders."

The brown tree snake, a voracious native of Australia, Indonesia, and Papua New Guinea, has almost completely wiped out the bird population in Guam, where it arrived in the wake of World War II, mostly likely by ship or plane.[36] And with almost no remaining birds, Guam's trees have lost their most valuable seed-spreaders. The result could be a reduction in new forest growth as high as 90 percent.[37]

The conquest of the United States by the potentially Zika-carrying Asian tiger mosquito has been traced to a single shipment of used tires from Japan in 1985. Within a decade of being discovered in Houston, Texas, it had spread to twenty-five states. Today it has been found in forty states, and has also made a home for itself throughout South and Central America, in Africa, and in the Middle East.[38]

The Ukrainian quagga mussel[39] was first spotted in the Great Lakes of North America in 1989. A single quagga can produce a million eggs a year, and within twenty years the quagga had spread from border to border and from coast to coast, devouring phytoplankton and disrupting the food web from the bottom up, devastating populations of salmon, whitefish, and native mussels.

Again and again we've learned that once a so-called invasion has begun, it's dreadfully hard to turn it back. We live in a world, though, with a cult-like adoration for the "never give in, never give in, never, never, never" part of Winston Churchill's famous World War II speech at Harrow School. And this sentiment has not failed to take hold of the world of ecological conservation. Perhaps not coincidentally, it was one of Churchill's contemporary Englishmen, the zoologist Charles Elton, who gave us the word "invader" to describe nonnative species in his 1958 book, *The Ecology of Invasions by Animals and Plants*.[40] Elton had lived through two world wars, and during the second, he had been tasked with controlling the pests that were infiltrating British food storage facilities, "and thus were practically in league with the Nazis."[41]

Elton, who later helped found the Nature Conservancy, fancied himself a "war correspondent" in a battle many people did not yet know they were in. He did not mask the militaristic underpinnings of his ideas, noting in the preface of his book that his first thoughts on plants and animals as invaders came during World War II. He warned "it is not just nuclear bombs and wars that threaten us."[42, 43]

Called to battle to "save" the world, we've spent billions of dollars to fight nonnative species in the United States alone. It's important to note that those costs are borne in an effort to cut back the trillions of dollars of damages these species have done to the American economy. The emerald ash borer, a native of Asia that arrived in North America in the 1990s, needed just twenty years to do more than $10 billion in damage to ash lumber crops.[44]

But are we throwing good money after bad? In many cases: Yes. Almost certainly.

What we sometimes forget in this never-say-die world is Churchill's caveat to the principle of steadfast resoluteness: "except to convictions of honor and good sense." In other words, you've got to know when you've been beaten. And especially when it comes to the smallest species in the world, we've been beaten.

The brown tree snake is a lot bigger than *K. huna*, but on a heavily forested island the snake has been nearly impossible to find and destroy. This has given rise to some bizarre solutions to the problem, like an airdrop of poisoned mice—parachutes and all—but so far, the snake abides.

And the smaller things get, the harder it is to even fathom a solution, and the more bizarre—and more frightening—many potential solutions start to sound. The National Wildlife Federation has concluded that the fight to eradicate invasive mussels is "impossible," with no solutions yet uncovered that don't grievously harm other wildlife.[45] And as for the tiger mosquito? Among the most promising proposals is the "release of insects with dominant lethality" or RIDL, which involves the mass release of genetically modified mosquitoes into the wild, a solution that some worry could turn the entire United States into Dr. Moreau's island.

And these are species we can easily see. *Kikiki* we can't. Not well, anyway. And that's to say nothing of the trillions of microorganisms that cross every border in the world, every day, on our hands and in our mouths and in our guts. Now that's bottom-up disruption.

This doesn't mean throwing in the towel. But it might mean throwing away old notions of what is "natural" and what is "invasive."

That's what ecologist Mark Davis suggested in 2011, when he took point in a company of nearly twenty other scientists to argue, in the journal *Nature*, that "the practical value of the native-versus-alien species dichotomy in conservation is declining, and even becoming counterproductive."[46]

Among many in a conservationist culture driven for decades by militaristic analogies—folks who had witnessed the damage voracious nonnatives can do—Davis might as well have announced himself as the reincarnation of Benedict Arnold. More than 140 scientists signed a sharply worded rebuttal to Davis and his faction.

But Davis was never suggesting that every nonnative species should be welcomed everywhere with open arms. His contention was simply that our response to nonnative species shouldn't rest on notions of "good versus evil" and that the world we've created—a world of trucks and ships and planes, and people moving from place to place, every day, carrying with them their microscopic menageries, not to mention things like increased urbanization, land-use changes, and, of course, climate change—had already inexorably globalized our ecological world. There simply is no going back.

As more of us come to recognize that the smallest battles of this war have already been won, our "convictions of honor and good sense" are almost certain to change.

And that's good. Because, over time, science should change the way we see the world. In ways big and small.

HOW THE WORLD'S SMALLEST VERTEBRATE COULD HELP US UNDERSTAND PAST CLIMATES

Remember how little we know about the world's largest frog? Well, we might know even less about the smallest, *Paedophryne amauensis*.

The world's tiniest known vertebrate, which was discovered in 2009 on the southeast peninsula of Papua New Guinea, is so small that it had long been missed—and not just by Western researchers. "The locals had no idea about these things either," the frog's co-discoverer, Louisiana State University herpetologist Christopher Austin, told me. "It's crazy because the locals are very, very good natural historians. They basically know all of the species in the forest."

But these frogs are teensy. At less than 8 millimeters from nose to tail, they're about the size of a kernel of corn. Being so small, they're really vulnerable to predation in a place with plenty of hungry animals—and even some carnivorous plants that are big enough to swallow up a tasty little amphibian. So they don't hang out on the exposed part of the rainforest floor, but rather *between* the dense layers of decomposing leaves that are matted all around the ground. It's a wonder they were found at all.

In fact, *P. amauensis* might never have found its way into the scientific record books to begin with were it not for a bit of serendipity. It was "a slow night" in New Guinea, Austin recalled. There are lots of nights like that, he said, when one is "herping." The frogs he and his team were actually looking for just weren't making a lot of noise, so the herpers didn't have a good place to start.

Without the distraction of the noises they were accustomed to hearing, Austin, along with his then-PhD student, Eric Rittmeyer, and their colleague from the Papua New Guinea National Museum, Bulisa Lova, took notice of a rhythmic, high-pitched noise that didn't sound much like a frog at all. To me, the call they recorded is reminiscent of the sound of a needle repeatedly bumping at the end of a vinyl record.[47] To them, the call sounded like the chirping of an insect. And while they were supposed to be looking for frogs, Austin said, "we were just a bit bored."

So off they went, stumbling around in the rainforest in the middle of the night, sending flashlight beams this way and that, trying to pinpoint the origin of the chirping.

And, well, they didn't find anything.

The noise was coming from something, of course. So Austin grabbed a big handful of leaves, stuffed them in a resealable plastic bag, and called it a night. The next morning he began slowly peeling the clump of leaves back, one by one. And there, looking back up at him, was an almost inconceivably small frog, dark brown with rusty-brown blotches and irregular bluish-white speckling. There are very few people in the world who know as much about frogs as Austin, and he'd never seen anything like it.

That frog, of course, set into motion a search for others. And when all was said and done, Austin and his team had collected enough specimens to conclude that, at an average length of 7.7 millimeters, they'd discovered the world's smallest vertebrate.[48]

While he's certainly proud of the discovery, Austin is circumspect about it, too. Could there be another, even smaller frog out there? "There might very well be," he said. "There's a bunch of new species of these micro-frogs that haven't yet been described, and it's quite possible that one of them is smaller than *amauensis*."

A search for an even smaller frog isn't just about setting another superlative record. We live in a world in which every new genome, and especially those belonging to organisms at the far edges of extreme physiology, can add to our collective understanding of our own genetic history. It's a world in which creatures that are evolutionarily distinct might offer us novel clues for curing diseases; the genus of New Guinean lizards called *Prasinohaema*, for instance, have bright-green, bile-pigmented blood that Austin suspects might hold clues for fighting malaria. It's a world in which every newly discovered frog species might help us get closer to a flu-free planet, since the mucus on the skin of some frogs has been shown to neutralize influenza.[49]

While a lot of the potential held by new species is pharmaceutical, it's also evolutionary and climatological. *P. amauensis*, which moved against the slowly and erratically waxing nature of Cope's Rule, didn't become an extreme species overnight. Genetic analysis shows that the miniaturized vertebrate, and several of its close and almost-as-small cousins, came along early into the evolutionary dispersal of frogs in New Guinea. We also know there's a big challenge to being a little frog: It has to stay moist all the time. Even an hour out of the water-saturated leaf litter could spell doom by desiccation. What these two things together tell us is that New Guinea's climate has been unfailingly stable for a very long time—always wet enough to ensure that these frogs had sufficient moisture to keep evolving smaller, but never so wet that the leaf litter flooded to the point of drowning them.

And now that we know that these frogs are there, so densely packed together that there may be thousands in a single acre of forest floor, we'll have another unfortunate indicator by which to observe and measure the impact of global warming in New Guinea.

They've also given us a whole new ecological niche to search. One of the main reasons we missed the existence of these little guys for so long is because researchers were often focused on the surfaces above and below—but not within—the leaf litter. Just as entomologists found *T. nana* after they began looking for *K. huna*, it's not unreasonable to believe that future expeditions to examine the leaf litter of New Guinea's wet forests could produce discoveries of not just more itsy-bitsy frogs, but other previously unknown creatures as well. Each new discovery will carry genetic, chemical, and evolutionary secrets that, once unlocked, could benefit us all.

It's not just the vast and largely unexplored forest floors of New Guinea we should be searching. The ocean, Austin reminded me, is even more vast and even more unexplored. He thinks the title of "smallest vertebrate" could very likely be claimed by a fish.

Some people think it has been.

Arriving at a concept like "smallest" isn't a purely objective endeavor. Is smallest a matter of weight? Of total mass? Of length? And, if it's a matter of length, then starting where and ending where?

Frogs are measured from "nose to vent," but their anus is on the extreme end of their backside, so an end-to-end measurement is rather easy. A similar measure of the previous record holder for "smallest vertebrate," a Southeast Asian fish known as *Paedocypris progenetica*, would cut that fish's length by half.[50] "Applying the same way of measuring frogs to *Paedocypris*—that is, using the same landmarks—clearly demonstrates that the fish *P. Progenetica* is still the smallest vertebrate," Ralf Britz, an ichthyologist who helped put the fish in the record books before it was supplanted by the tiny frog, told Phys.org in 2012.

Austin—an enthusiastic angler—says no one in their right mind would measure a fish that way.

The debate over the smallest known vertebrate—such that it was even a debate—was always a good-natured one, Austin said. But it also highlights yet another reason why researchers tend to avoid applying superlative descriptions to their discoveries, even when those discoveries are, in fact, exceptional. While good science is often controversial, few scientists actively court controversy. And in any case, Austin said, there's plenty of research that can and should be done on both species—and more like them—regardless of which is presently deemed to be the record-setting smallest.

On the other hand, a little controversy isn't bad for publicity. And the difference between a paper describing a very small vertebrate and the smallest-known vertebrate is night and day when it comes to getting people to pay attention. While plenty of scientists would sooner eat lunch from a well-used Petri dish than talk to the public, the need for more science in the media is clear.

To bring attention to his discovery, Austin did something brilliant. Almost immediately upon discovering *P. amauensis*, he fished a dime from his pocket and set the little frog upon it. The photo that resulted has been shared millions of times over, all around the world.

There was one thing that Austin could have done better, though, when it comes to public relations for his research.

"So, um . . . Pa-ey-do, um, fa-rene, um, amnu, amen, am . . . oh heck, man, I don't know how to say this," I told Austin when we first spoke. "Do you have a nickname for them?"

"We call them micro-frogs," he said.

"I like that much better," I replied.

If we want people to feel an affinity to nature, names are an important place to start. *Tinkerbella nana* is a great example of this. Even if you haven't seen it, you sort of can't help but love it.

There is at least one species that seems perfectly designed for great PR. With a superlative claim to adorable fame, a bit of good-natured debate, and an endearing name, this little guy could be one of the best potential science ambassadors the world has ever known.

HOW A TINY BAT HELPED SOLVE AN EVOLUTIONARY MYSTERY

Let's start with the name. It actually has a few of them. Its scientific name is *Craseonycteris thonglongyai*. It's also known as Kitti's hog-nosed bat. For my money, though, its third name—the bumblebee bat—is a perfect moniker, because this Southeast Asian bat could, at a glance, quite easily be mistaken for a black bumblebee.

And yes, it's as cute as you might imagine, with a fluffy mane of gray or sometimes light brown hair, a pink face, a perpetual smile, and ears shaped like butterfly wings.

It's also the last known remaining member of the Craseonycteridae family of bats, which diverged from other bat species about 43 million years ago. That gives it peerless potential as a source of information about the genetic history of its cousins in the order Chiroptera—animals that many scientists believe may be facing a little-recognized extinction crisis, and which others note are a valuable indicator species for woodland and wetland environments worldwide. The more we know about bats, the better they'll help us understand how habitat destruction and climate change are impacting our world as a whole.

Like a lot of superlative species, it wasn't until quite recently that we even knew the bumblebee bat existed. It was only discovered in

1974, and for nearly two decades afterward its only known habitats were a few caves in western Thailand's Kanchanaburi Province.

Then, in 2001, taxonomist Paul Bates was taking a census in a limestone cave in southern Myanmar, perched 50 feet from the cavern floor with a net covering the entrance. A tiny bat landed in the net, inches from his right hand. Bates grabbed it, and as soon as he did he suspected that he'd found something amazing.

"I think I know what you are," he told the creature, "because you're so damn small."

Later that day, when he gently lifted the tiny bat from his collection bag and examined it, his suspicion was confirmed. "I always had it in my mind that they might be there but then, when you finally see it, it's just very exciting," he told me. "I remember thinking, 'This is a good day's work.'"

That good day led to good years of further research, thanks in large part to the bat's superlatively small size. Bates said officials from Myanmar's scientific community were thrilled to be able to stake a partial claim to the world's smallest mammal, opening up opportunities for scientists from around the world to work with colleagues in one of the world's most restrictive countries.[51]

Being able to add the words "world's smallest" to grant applications was a bonus for researchers, too.

"I've sat on grant application boards before. It's quite tedious, to be honest," Bates said. "And a lot of the applications, especially if they're highly technical and not in your field, you think, 'Oh, not another one.' But if there's something like a hook, and they use the hook well, it certainly helps," he said. "This bat weighs two grams and it does all this really clever stuff. It's not difficult to sell it to grant givers and to the general public."

None of that makes the science any less compelling. "It's cute and it's interesting, but it's also interesting scientifically," Bates added. "This bat is the last of its line . . . it's kind of successful and unsuccessful at the same time. It's very old and doesn't appear to have evolved that much, but yet it had a tiny distribution. It asks a lot of questions."

Bates soon found himself on an international team of researchers headed back to Southeast Asia to explore a chicken-and-egg riddle of evolution.

The bumblebee bats found in Myanmar were very similar to the ones discovered in Thailand. But the researchers noticed a curious distinction. The calls of the Myanmar and Thai populations were as much as 10 kilohertz apart in frequency. Bats are known to slightly fluctuate their calls to prevent confusion when in proximity of other bats, much like referees on adjoining soccer fields sometimes use differently pitched whistles to avoid accidentally stopping the wrong game. But the differences between the Myanmar and Thai bats wasn't slight. If these two populations were on the AM radio dial, they'd be stations for classic rock and country.

Because the bumblebee bat is evolutionarily distinct, with no genetic in-flow from other closely related species and subspecies, it was a perfect animal by which to examine a long-popular hypothesis: that differences in mating traits—a bat's calls, for instance—can drive an animal in two different evolutionary directions, resulting over time in the evolution of one or more new species.

The challenge of researching this "sensory drive hypothesis" is that it's hard to determine whether a variation in sensory traits was a cause of evolution, or the effect of another evolutionary driver. Did differences in the calls of these two groups of bumblebee bats drive them to live in separate geographic locations, for instance, or did the geographic separation prompt the changes to the calls? Using genetic samples from both populations of the bumblebee bat, Bates's team went searching for DNA-level differences.

Even after an estimated 268,000 to 545,000 years spent apart, the genes of the Thai and Myanmar bats were remarkably similar—a testament to their fitness for life in the 2,000-square-mile area of Southeast Asia where they are found.

There were, however, significant differences in a gene called *RBP-J*, which has been shown to be involved in inner-ear hair cell formation—an important part of hearing for humans and bats alike.

By measuring other, smaller differences in the genomes of sub-sub groups of the bumblebee bat, the team concluded *RBP-J* mutations must have come after the Thai–Myanmar split—meaning the need to adjust to differing environments drove a shift in sensory adaptation, and not the other way around.[52]

That's not a death knell to the sensory drive hypothesis, but it does provide a key piece of evidence that has already informed dozens of other studies and research efforts tied to that postulation. Today we have a more nuanced understanding of the drivers of evolution, thanks to one of the world's smallest mammals.

Why do I say "one of the world's smallest"? Because just as there is some debate over whether the micro-frogs of Papua New Guinea can rightfully claim the title of "world's smallest vertebrate," the bumblebee bat does not hold the undisputed title of the world's smallest mammal.

HOW A LITTLE BLIND SHREW IS HELPING US SEE THE WORLD IN NEW WAYS

It's true that the thumb-sized bumblebee bat is the shortest known extant mammal. But the slightly longer Etruscan shrew, which weighs in at 1.8 grams, is slightly lighter.

Is it hard to imagine what 1.8 grams feel like? Go grab a sugar packet, and empty about half of it into your hand. That's what it feels like to hold *Suncus etruscus.*

Like the bumblebee bat, the Etruscan shrew is ripe with potential for research. And, as with so many other species of superlative distinction, there are a lot of questions that haven't been answered, or even asked, because we're still gathering some really basic information—like where it lives.

Part of the problem is that the shrew may be the animal kingdom's greatest escapologist, too light to set off many weight-triggered traps, small enough to slip through the thinnest of openings in a cage, and so fast that it's simply tough to catch in the first place. What we do know of their distribution often comes from owl pellets—the undigested bones, teeth, and hair owls regurgitate and leave on the ground as evidence of their diets for ornithologists and elementary school scientists alike. But there are plenty of places where owls don't roam, where they decide not to bother with such a small snack, or where they're simply not very successful at catching incredibly small and fast prey.[53]

When it comes to understanding the way our brains work, these hard-to-catch critters could offer crucial clues. The shrew's cerebral cortex—an area of the brain that plays vital roles in memory, attention, and perception—appears to share many common aspects of brain circuitry with other mammals, including us. But because it's so small, it may help us solve a challenge that has vexed brain researchers for years: getting a good image of the cerebral cortex at work.

For many years, neurobiologists have used two-photon microscopy, a process by which near-infrared light is used to "excite" fluorescent dyes to produce a brilliantly detailed multilayered image of living tissue. The technique is generally only good for tissues up to 1 millimeter thick, though. In most mammals that's enough to get a picture of only a few layers of cerebral cortex cells. But the Etruscan shrew's cerebral cortex is thinner than a credit card—and, as a result, researchers can record activity from entire sections of its brain at once.

That might not mean much to most folks, but it was a game changer for Robert Naumann, an expert in neural computation from Humboldt University in Berlin. He's suggested the Etruscan shrew might be the perfect model for understanding the relationship between the structure and function of different parts of the brain, even allowing scientists to record the activity of every neuron in an area like the visual cortex—a development that could tell us how mammalian brains process images collected by the eyes.[54]

S. etruscus is also helping us "see" in other ways because, like many burrowing animals, it doesn't have the best eyesight in the world. It can nonetheless identify potential prey—a cricket, for example—and decide whether or not to attack it in less than 30 milliseconds, based on nothing more than how the insect felt when it brushed up against the shrew's whiskers.[55] To do that, the shrew's brain has to process an "image" of the prey created by its whiskers as fast as, or even faster than, human brains process the pictures collected by our eyes.

For Tony Prescott, a neuroscientist from the University of Sheffield's Center for Robotics, that wasn't just a biological feat to marvel at; it was a robotic challenge.

Most robots that interact with the world do so via some sort of interface reliant on the visual spectrum, from simple cameras to laser beams. Prescott was interested in identifying other ways in which robots could get a glimpse of the world around them, and when he learned about the shrew's uncanny ability to rapidly identify prey from whisker touch, he began to hatch a plan. The result was Shrewbot, a camera-less robot that can map its surroundings based on the movement of magnets at the roots of its eighteen whiskers. In the future, whiskered robots could operate in places where other sensors don't work very well—like smoke-filled rooms, clogged pipes, planets with thick atmospheres, or deep ocean trenches where using light could harm animal life.[56]

Could something like Shrewbot even have been conceived if not for the existence of an actual animal whose behavior it dictates? The best we can say, of course, is "perhaps," for scientists and engineers are endlessly clever folk. What can be said with reasonable certainty, though, is that the Etruscan shrew itself could not have developed the traits that inspired the robot were it not for its incredibly small size, which allows for stimulus-driven signals to move exceptionally fast from the whiskers to the brain to the rest of the body.

What, then, might be inspired by a closer examination of the world's smallest known mollusk, a .33-millimeter water snail from Cuba called *Ammonicera minortalis*? What might we learn from the smallest known beetle, a similarly sized creature from South America known as *Scydosella musawasensis*? What could come from a deeper examination of the smallest known reptile, *Sphaerodactylus ariasae*, or from the smallest bird, *Mellisuga helenae*, or from the smallest primate, *Microcebus berthae*? None of these creatures has been particularly well examined.

What could we know if we spent even a little more time and effort looking for the smallest things in our world?

When I put that question to Ken Williams, the Colorado research site manager who had helped introduce the world to uranium-breathing microorganisms, he noted that the Rifle team wasn't even initially looking for ultra-small organisms. Rather, he reminded me, it was seeking to answer questions about how bacteria could be used to neutralize radioactive waste.

It wasn't the research question that led to a significant rewriting of the tree of life. It was a happy accident. "The most important thing, in my view, is creating the kinds of conditions that allow you to be lucky," he told me.

And at Rifle, he said, that was all about assembling a multidisciplinary team of brilliant minds around one question while encouraging everyone to ask and seek to answer broader questions, as well.

What might come of assembling a similar scientific Justice League around the exploration of the smallest of things in our world? Already we have a pretty good idea: new understandings about how evolution works, innovations in gene sequencing, advances in human health, and discoveries of entirely new environmental niches.

And that's no small thing.

Chapter III

THE OLD DOMINION

How Our Biological Elders Are Offering Us New Knowledge

The drive west out of Rifle on Interstate 70 is a study in Western panoramas. The freeway chases the Colorado River for about 80 miles before parting ways at the Utah border. From there the river breaks left, along the southern edge of Arches National Park, cutting a path through the middle of Canyonlands before proceeding south to the Grand Canyon.

I longed to follow the great river, but the road ahead of me offered a natural treasure just as rich. I was going back to Pando.

Forest ecologist Burton Barnes hadn't just made a spectacularly good guess as to the Fish Lake aspen clone's immense size. He had also proffered an estimate of its age. By comparing the distinct appearance of the clone's leaves to similar-looking fossils, Barnes concluded it could be as old as 800,000 years.

Cope's Rule specifically refers to evolution of increasing size, but the basic principle could be applied to any attribute of life. There's always room at the top, after all. Room to grow heavier. Room to grow taller. Room to grow older. And as something grows in one way, it often grows in others.

If Barnes's estimate about Pando's age is true, it wouldn't just be the oldest known organism in the world, but the oldest by a spectacular factor—perhaps even the oldest that ever lived. By Barnes's reckoning, the great plant's birth may have come alongside some of our earliest human ancestors: the same time *Homo erectus* began to harness fire and our forebears' brains began a rapid evolution in size and complexity.[1]

Human DNA has changed significantly in the millennia that have come and gone. But the essential code written when Pando was created—the very record of that time and place as expressed in an aspen genome—remains as it ever was, such that running your fingernails along its chalky white bark puts you into visceral contact with life not only as it is, but as it has been, perhaps for as long as *we* have been.

As was the case with Barnes's estimate about Pando's incredible size, though, plenty of people have expressed doubts about his guess on its age. But while we now know, with a significant degree of certainty, how far Pando's genes have spread across the Fish Lake Basin in terms of acreage, its age has proved harder to confirm.

The age of most trees is relatively easy to determine, and just about everyone knows how to do it. It starts with a clean cut across the trunk. Once the rings are exposed, the counting begins. With some scientific caveats, each ring represents a year.

Clonal stems can be aged in this way, too. I haven't cut down any of Pando's stems, but there are plenty, fallen in the natural way, lying about on the ground, and forest managers have cut many of these into easily countable cross-sections. One I counted, on a cold fall afternoon, had 87 rings. Another had 94. Another still had 103. But none of the stems are as old as the clone itself; not even close. Like the hair on a person's head, new ones emerge as older ones die, but the forest remains.

Some have concluded this somehow disqualifies clonal organisms as life-forms worthy of a superlative description like "oldest." I disagree. An aspen clone is not unlike you and me. When we look at ourselves in a mirror, we might recognize our face as the face we've always had, but it's not, and all it takes is a look at a photo from ten years earlier to be reminded of that fact. The cells that make up our skin last a few weeks, at best. The cells in the rest of our body die and regenerate at rates varying from every few days to every few decades. You likely still have some living cells that were with you at your birth, but not many, and each day there are fewer. Over time, as our cells come and go, our appearance changes. So no, you are not the you that you were—but you are still you.

You, however, most likely have a record of your birth. Pando doesn't. So how can we be sure it's so old? The short answer is that we can't be. Not yet, anyway.

Some have reasoned that the maximum known growth rate of aspen clones can be divided into a clone's total size to give a minimum possible age. Using such a formula, some researchers have guessed that Pando might be about 80,000 years old, and in recent years popular opinion has coalesced around that figure.[2] That's a far cry from Barnes's estimate, but even at 80,000, Pando would still be the oldest living organism we know of by a significant factor, and its birth would have come around the same time that humans began to take their first steps out of Africa.

However, there are clear limitations to such an estimate. Aspen exhibit tremendous genetic diversity and live in a lot of different habitats. And just as humans carrying different genes and living under differing circumstances grow in different ways—Dutch men are six feet tall on average, for instance, while their average Indonesian counterpart is 5-foot-2[3]—we can safely assume that aspen also will grow to different sizes under different genetic and ecological circumstances. Pando could be an amazingly fast-growing specimen; 80,000 years could be liberal. But there's also the matter of the fires, floods, landslides, and herbivores that, over thousands of years, could have taken big bites from the clone's size. Pando might actually be quite a bit smaller now than it once was; 80,000 years could be conservative.

Soon, we might have a better idea.

When I met him in 2018, Jesse Morris, who works at the University of Utah's Records of Environment and Disturbance Lab,[4] told me he was hoping to secure grant funding to take sediment cores from the middle of Fish Lake.

Most lakes in the region were formed by glaciers, Morris explained, and are somewhere in the neighborhood of 8,500 years old. Fish Lake, though, was formed tectonically as the Great Basin pulled away from the Colorado Plateau. "That lake might be a million years old," Morris said.

And Morris estimates that under Fish Lake, which is more than 100 feet deep in places, there may be 30 or 40 meters of compacted sediment—a gold mine of evidence about past climates and, potentially, what plants lived nearby tens of thousands of years ago. "It's actually pretty fortuitous, and a really unique opportunity, to have such a long-lived organism next to such a mind-bogglingly old lake," he said. "You couldn't ask for a better situation."

The lab where Morris works has had some success using sedimentary pollen fossils to reconstruct climates that existed thousands of years ago. Using the pollen fossils, the team has even demonstrated the ways in which aspen in the Rocky Mountains moved upslope during a period of drought more than 4,000 years ago.[5]

Morris doesn't know if it would be possible to tie exceptionally old pollen to a *specific* aspen clone. But if aspen pollen disappeared altogether from the samples as the researchers moved deeper and further back into time, it would give them a much more definitive "oldest possible" age.

Knowing Pando's age with greater certainty wouldn't just give a few in-the-know folks another superlative fact to pull out on bar trivia night. It would help all of us understand the upward limit for how long a specific genome can exist in our world. It would tell us something about the potential longevity of the fundamental building blocks of life on our planet. And it could tell us something about ourselves.

We've long known how unique human genomes can help us unlock secrets about our lives. Even before scientists knew what a genome was, at least some of them understood that humans with rare genetic conditions could be a blessing to the rest of us.[6] As far back as 1882, the British pathologist James Paget recognized that treating people with rare conditions as "freaks," as we all too often did, was a scientifically, as well as morally, bankrupt approach. People who are very different shouldn't be set aside "with idle thoughts or idle words about 'curiosities' or 'chances,'" he wrote. "Not one of them is without meaning. Not one that might not become the beginning of excellent knowledge, if only we could answer the questions—why is it rare? Or being rare, why did it in this instance happen?"

This is also the question cancer researcher Josh Schiffman was trying to answer when he broadened his research focus to include not just the experiences of people, but tragically cancer-prone dogs and providentially cancer-free elephants, too. The result of that inquiry, as we know, has fundamentally shifted a lot of thinking about how to fight cancer.

And this is why the question of Pando's age is so important. Plants aren't people, but we share a common eukaryotic ancestor, which means we share quite a few genes. And of course, aspen are even more closely related to other plants. The strategies Pando has used to stay fit for this world, millennium after millennium, could inform our own efforts to survive and thrive on this quickly changing planet, particularly when it comes to ensuring the survival of other plants.

So it would indeed be shameful to set aside this organism's longevity as a mere curiosity. For it might also be the source of excellent knowledge, if only we seek to answer the questions "Why is it old?" and "Being old, how did it get that way?"

HOW STERILITY CAN HELP AN ORGANISM GROW OLD

The genetic research Karen Mock did on aspen didn't just help substantiate Pando's claim to fame as the world's largest known organism. It also revealed a rather big genetic curiosity: A surprising number of aspen, Pando among them, had three sets of chromosomes.

Most eukaryotic species have two sets. And a lot of aspen are diploids, too (as are we). But when Mock and her team studied aspen across North America, they found that up to two-thirds of the plants in some regions were triploids. That came as a bit of a surprise, because everything we know about biology tells us that triploids typically have one heck of a hard time reproducing, since their cells can't properly undergo division.

Research from two other species—both also clonal—is helping put Pando's sterility into context.

The first is another contender for the title of "world's oldest plant," the last known member of the species *Lomatia tasmanica*, also known as King's holly. The plant was first identified by naturalist Denny King in 1934, but it wasn't until 1998 that carbon dating of identical-looking fossil leaves found nearby revealed the plant might be 43,000 years old—and perhaps even older.[7] While the plant does produce pink flowers, it doesn't produce fruit or seeds because, just like Pando, it's a triploid.

Across the Bass Strait is another part of the puzzle, *Grevillea renwickiana*. There are fewer than a dozen individual specimens of *G. renwickiana* left in the world, all of them in the plant's native Southeast Australia. And, like Pando and *L. tasmanica*, the plants are triploids. Sterile as a mule.[8]

While there aren't many *G. renwickiana* left, the survivors are doing just fine and dandy. One, in fact, is spiderwebbed across an area near the Endrick River, in Morton National Park, that stretches out for nearly 82 acres. That's not much less than the territory claimed by Pando.[9]

Mass sterility ought to spell the end of a species. But conservation geneticist Elizabeth James, perhaps the world's leading expert on *G. renwickiana*, told me it probably makes a lot of sense that these three plants are survivors. "If they're sterile, they don't spend energy on reproduction," she said.

"But if they don't spend energy on reproduction won't they eventually die off?" I asked.

"They are certainly evolutionarily stagnant," she said, "but the ones that are left seem to be flourishing."

Instead of creating lots of flowers and seeds—which are the most complex parts of a plant and take a lot of energy to produce—the triploid plants can dedicate available water, sunlight, and nutrients to building robust root systems. "That could give them an edge," she said.

I don't know a parent who can't relate. Who among us doesn't feel as though their children have taken years off their lives?

Jokes aside, though, there is actually a lot of evidence of an inverse correlation between fertility and longevity. In nearly every organism that has been studied over the past half century, in fact, reproduction shortens lifespan.[10] And when researchers from Moscow State Univesity investigated the relationship between lifespan and fecundity in 153 countries around the world, they found a highly significant negative trend; after controlling for religion, geography, socioeconomic factors, and considerations of disease, they still observed a trade-off between the average number of children and life expectancy.[11]

Those trade-offs are scientifically significant but relatively slight, though. What more likely accounts for the apparently exponential longevity of plants like *P. tremuloides*, *G. renwickiana*, and *L. tasmanica* is their interconnected root structures. Surface soil protects them from short-term environmental shifts that could kill plants reliant on sexual reproduction.

"If you have diploids producing seeds," James said, "and the environmental conditions change and the seeds don't get to germinate, then an entire generation can disappear. Once that individual dies, it's gone, whereas the triploid ones can get going again."

To see where this sort of knowledge might come in handy, it's helpful to think about the sort of world scientists believe is coming as a result of human-caused climate change. There's virtually no doubt the planet is getting hotter. What we also are coming to understand is that weather extremes—from hurricanes to heat waves to historic droughts—are a part of the price we're paying for unleashing vast quantities of greenhouse gases into our atmosphere.

What plants are most likely to survive such disturbances, and even mitigate the resulting ecological instability? Perhaps, James suggests, it will be those like *P. tremuloides*, *L. tasmanica*, and *G. renwickiana*, which have been doing so for millennia without the need for renewal by germination.

It's difficult to know with certainty how old the last surviving *G.*

renwickiana clones might be. "But they're surely really old," James told me. The diploid version of *G. renwickiana* may have died out many thousands of years ago, she said. The triploids, though, may be around for thousands of years to come, even in the face of climate change.

If we don't kill them off in other ways, that is. Because, let's face it, humans have a knack for bringing an end to things that Mother Nature has kept going for a very long time.

HOW ANCIENT ASPEN ARE TEACHING US ABOUT INTERCONNECTEDNESS

Eight hundred thousand years? Eighty thousand years? Whatever we learn about Pando's age, this much is certain: The aspen clone was around long before white settlers colonized this area of North America in the 1800s.

Will it outlast us? That is far less certain. Because this amazing organism, which might very well be the oldest living thing on our planet, now appears to be dying.

When wildland resources researcher Paul Rogers first took a walk through Pando in 2010, he was unsettled by what he saw. It was as though the plotline from P. D. James' dystopian novel, *The Children of Men*, had been rewritten with "Pando" in the place of "people." In James' 1992 book, and *Children of Men*, the 2006 film adaptation, human children have inexplicably stopped being born, and the aging population that remains on Earth has been thrown into chaos.

Pando still had many "geriatric" stems, some 100 years old and likely even older. There were plenty of "senior citizens" as well—those in their seventies and eighties. There were very few young adults, though, and just a smattering of teenagers. And there were almost no children. The new stems had simply disappeared.

"Something has been disrupted," Rogers told me as we hiked through an area of the clone that had been particularly hard hit, where recently fallen stems covered the ground like pick-up sticks, leaving nothing behind but a clear view of Fish Lake. "There really hasn't been any new growth for three, four, five decades."

Even more alarming to Rogers was that few people had so much as taken notice. This great and ancient thing was tumbling off Cope's Cliff, and no one was doing anything about it.[12]

That might be the right course of action if Pando had simply reached the end of its remarkably long life. But that's not likely what was happening.

"The evidence certainly suggests that something has changed in the recent history of this plant," Rogers said. And identifying the assailant in this mystery wasn't tough. The Fish Lake clone was humming along, alive and well, for millennia, and then *we* showed up.

But like a game of Clue, it's not enough to know the culprit. We've also got to figure out what the weapon was.

One guess: fire. Or rather the lack of it. Aspen thrive in the midst of disturbance. Cut down a clone's stems, and a healthy aspen root system will send up many more in replacement. Allow a fire to rage through a colony, and new suckers will often follow the path of the flames. But Fish Lake is a popular recreation area, and with cabins scattered throughout its forests—including a few dwellings that are actually situated within Pando's boundaries—state fire managers hadn't sat idle. The fires, which once, quite naturally, came through here every now and again, have been stopped.

Another potential weapon: climate change. One joint research effort from the US and Canadian forest services demonstrated that regions impacted by drastic aspen die-offs are those that have experienced hotter temperatures and drier winters in recent years.

Rogers had yet another theory: deer and elk. Too many of them. It had been some eighty years since the region's chief predator, the gray wolf, had been exterminated. That timeline roughly coincides with the last big generation of stems in Pando. The area's mountain lion population had also been drastically reduced in the past century, first by a bounty program that claimed nearly 4,000 cats between 1913 and 1959, and later by game regulations that permitted hunters to take any number of big cats at any time without a permit.[13]

To understand the impact of predators on aspen ecosystems, it's helpful to head north from Pando—just about 350 miles, as the crow flies—to the Crystal Creek area of Yellowstone National Park.

The last of Yellowstone's endemic gray wolves was exterminated in 1926. Then, in the mid 1990s, thirty-one wolves were reintroduced to the park, and it didn't take long to see their impact. There were 18,000 elk in Yellowstone at that time. They were always hungry, and one of their favorite snacks were the leaves from young aspen stems. But the wolves were always hungry, too, and one of their favorite snacks were the elk. When the wolves started doing what wolves do so well, the elk were no longer able to stay in one place for long periods of time and munch through entire groves of aspen. Soon, Yellowstone's aspen woods, like those in Crystal Creek, were flourishing.

Wildlife ecologist Dan McNulty, who has studied the wolves-to-trees phenomenon, said it could be many more years before it's clear whether Yellowstone's aspen have been saved. "The outlook for aspen in Yellowstone is promising, but not assured," he told me.

One thing was clear, though: The wolves' impact didn't stop at reduced elk and increased aspen. The ripples kept going and going. The bigger, healthier aspen woods offered habitat to birds and building materials for beavers, whose dams help raise the water table, thus providing habitat for even more trees.[14]

Rogers couldn't unilaterally reintroduce predators into Utah—there are a lot of hunters, farmers, and ranchers in that ultra-conservative state who would have something to say about that—but he wanted to know what Pando would look like if the local ungulates couldn't use that great and ancient organism as an all-you-can-eat salad bar.

"There's really nothing at all to keep them from bedding down in an area and just eating for a week or a month," he told me during a hike in the woods one day, "if they find a particularly tasty aspen."

I reached for the nearest stem, leaping to grab an aspen leaf. I popped it into my mouth like a piece of fresh spinach. It tasted like chewed-up aspirin.

"This is *not* a particularly tasty aspen," I said, trying to spit and finding the leaf had robbed me of my saliva.

"You're not an ungulate," Rogers laughed. "And besides, they tend to like the young leaves."

"Unless there is a wolf chasing them?" I asked.

"Right," he said. "But since we don't have wolves, we built a fence."

He led me to the gate of a fenced-off quadrant of the clone, an area known as the "research restoration area." The first two suspects in Pando's slow death, fire and climate change, would be hard to measure, particularly in the short term. But the third, the impact of deer and elk, could be controlled. Hence the fence—which at that point was just a month old.

It was Rogers' first trip to Pando since that fence had gone up. He wasn't expecting to see a big difference in that time. Even if the ungulates were part of the problem, he told me, fire and climate change, and "God only knows what else," were almost certain to be accomplices. Fixing one thing alone wasn't likely to save Pando.

But then he saw it: A tiny sucker stem, seven inches at most, peeking out of the ground near a rock, with just a few bright green leaves jutting from the top.

And then, near a fallen log, he saw another.

His amble picked up pace. Soon he was darting about the forest, this way and that, jumping over fallen stems, spinning around to find sucker after sucker after sucker, running his fingers over the leaves of shin-high shoots.

"Here's one," he called out. "And here's another . . . and another!"

He didn't look much like a scientist in that moment. He was more like a boy at play in a hundred-acre wood. By the time the sun began to set, though, he'd reclaimed a more composed disposition. "It's way too early to say what this means," he told me. "It's promising, for sure, but we need to wait to see what happens."

I didn't fully understand Rogers's excitement on that day. But five years later, I was back in the clone, walking the fence line with what I can only imagine was the same sort of oh-my-God-this-really-worked glee that Rogers had been trying to contain on the day he introduced me to Pando.

Inside the fence, the tiny shoots we'd spotted years earlier were sturdy stems now, and the ground was brimming with more first-year suckers.

And outside the fence? Almost nothing.

It's still too early to say whether Pando's revival within the fence will lead to the sort of resurgence of biodiversity seen in Yellowstone. But it's not debatable that over-grazing by ungulates has been a key factor—if not *the* key factor—in the Fish Lake clone's decline.

It would be expensive, albeit feasible, to build a fence around the entirety of Pando's perimeter. We could, I suppose, save this ancient thing that way. But that is not a scalable solution. We cannot build a fence around every aspen clone.

We could, however, listen to the voice of the father of wildlife management, Aldo Leopold.

"I have lived to see state after state extirpate its wolves," Leopold wrote in 1945. "I have watched the face of many a newly wolfless mountain, and seen the south-facing slopes wrinkle with a maze of new deer trails. I have seen every edible bush and seedling browsed, first to anaemic desuetude, and then to death. I have seen every edible tree defoliated to the height of a saddlehorn. Such a mountain looks as if someone had given God a new pruning shears, and forbidden Him all other exercise. In the end the starved bones of the hoped-for deer herd, dead of its own too-much, bleach with the bones of the dead sage, or molder under the high-lined junipers."[15]

Inherent in Leopold's essay, called "Thinking Like a Mountain," was a simple plea: Allow predators to play their rightful role in the ecosystems in which they evolved.

If we do, we might save some of the oldest living things in our world. And we'll most certainly save more than that.

WHY BRISTLECONE PINES ARE TAKING THE HIGHER GROUND

I had what amounted to a treasure map. A few clues about pinecone washes and switchback trails. A basic description of where to look and what to look for. And a very old photo.

But there were thousands of bristlecones along the trail. Twisted and gnarled, grasping and desperate, still against the stark blue sky. They looked like demons frozen while trying to escape hell. And the idea of finding one specific tree, among all of those in the Ancient Bristlecone Pine Forest in California's rugged White Mountains, seemed daunting at best.

But I kept hiking, searching for the clues I'd been given. And then I saw it.

Methuselah.

At 4,850 years old, it is the oldest known single tree on our planet, germinated at the same time that the first pyramid of Egypt was built. Only a few people know where Methuselah is, and they have been protecting its location like the Ark of the Covenant.

Here is why: A few decades back, when Methuselah was marked with a sign, visitors were taking samples of it home with them. They were, forest officials worried, going to "love it to death." So the park rangers took down the sign, resolving to give the world's oldest tree a shot at continuing to be the world's oldest tree.

There was an interesting duality at play in that decision. On the one hand, researchers who study bristlecone, like the University of Arizona's Chris Baisan, say the age of individual trees—even superlative ones—isn't scientifically important. "If you're a scientist rather than a trophy hunter," Baisan said in 2015, "you don't need the oldest individual."[16] On the other hand, tree-ring researchers and forest officials have gone to great lengths to protect Methuselah, which doesn't make a lot of sense if Methuselah itself really doesn't matter.

It was around the time Methuselah's sign came down that a dendrochronologist named Tom Harlan began telling people he had identified a bristlecone that was even older. No one who knew Harlan seems to believe he would just make up such a claim, but the longtime University of Arizona researcher declined to publish his findings, and took the secret of the tree's location to his grave in 2013.[17] Harlan's colleagues looked through his notes and collection of core samples for clues about the mystery tree. There was no sign of it.

It's possible, I suppose, that Harlan was spinning a tale to give Methuselah a bit more breathing room. But Harlan's colleague at the

University of Arizona, Matthew Salzer, said that if the tree does exist, he thinks he knows the area where it might be. He has considered going to find it, and I've begged him to take me if he does.

And yet I'm not sure I want to know for certain. There's something sacred about walking through the White Mountains without knowing which, among all of the ancients, is the holiest of holies.

The first time I hiked the Methuselah Trail, I didn't know which of the trees was oldest. And so my awe was incited on that day not by age, but by the relationship between age and size. Having witnessed the way thousands of years of growth could turn an aspen clone into an organism of godlike proportions, one of the first things that struck me about the ancient bristlecones in the White Mountains was just how very small they seemed. It would have taken mere seconds to climb their gnarled trunks and reach their upper branches.

Like most trees, bristlecones add a ring for each year of their lives, and some of these had been doing so since before Stonehenge was constructed. Each year's growth, though, comes in mere fractions of a millimeter—it is nearly impossible to accurately count the rings of a very old bristlecone with the naked eye.

It makes sense that the bristlecones of the White Mountains grow so slowly, Salzer told me when we met up, early one fall semester, in his lab at the University of Arizona, the same school where Harlan once worked. Bristlecones, Salzer said, live in some of the most inhospitable conditions imaginable, especially those at the highest reaches of the treeline, 11,000 feet above sea level, where temperatures stay below freezing for long stretches of the year, and where it can be extremely dry during the short growing season. "So they just take their time," Salzer said.

Only a few living bristlecones reach ages approaching Methuselah. But there is no shortage of dead bristlecones scattered about in the White Mountains, and some have been lying in state for thousands of years. By matching and overlapping the ring growth patterns from both living and dead trees—not unlike how the scientists in Rifle, Colorado, used overlapping segments of DNA to piece together longer sequences—Salzer is close to completing a 10,000-year timeline of bristlecone growth. "There are just a few puzzle pieces left," he said. "We just have to find the right pieces."

Matching the rings is both art and science. Elevation, slope, soil, and other factors can impact how individual trees grow from year to year, even within short distances of one another. So dendrochronologists search the rings for signs of climate events massive enough that they impacted every tree alive at a certain moment in time. Looking through a microscope in his lab, Salzer pointed out one such section in a bristlecone core sample that, to my untrained eyes, seemed "fuzzier" than the others. "That's a frost ring," Salzer told me. "Those coincide with large volcanic eruptions that sent a layer of dust across the world, and they're widespread—you'll see them in bristlecones from all over."

The frost ring I was looking at marked the year 627 AD, about the same time Muhammad conquered Mecca. Salzer and other tree ring researchers have found similar aberrant rings in bristlecone pines from 536 AD—when "the sun gave forth its light without brightness," as the early Byzantine historian Procopius wrote of the global "dust veil" that led to widespread crop failures and famine[18]—as well as 687, 899, 1201, 1458, 1602, 1641 and 1681 AD.[19] While the 10,000-year timeline is still coming together, such markers have enabled Salzer and his team to develop a robust data set exposing a 4,650-year timeline of bristlecone growth, and demonstrating the tree's slow-and-steady strategy for a long, healthy life. Over more than four millennia, the median annual ring growth was less than .4 millimeters, about the thickness of a human fingernail.

But something has shifted, recently, in the highest altitudes of the White Mountains. There, as of late, the bristlecones have been growing like gangbusters. Or relative gangbusters, at least. From 1951 to 2000, their growth averaged .58 millimeters—a mark that stands as a record within the timeline, and which had only even been approached once before.[20]

There could be a lot of explanations for this phenomenon, but when Salzer's team looked at the growth of rings during the period in which we have reliable temperature data, they saw a strong correlation between annual mean local temperatures and ring growth. This work is offering us much-needed context for the drastic climate changes we're now seeing in our world, and further proof—not that we should need it at this point—that these changes aren't simply part of some larger, longer cycle.

Bristlecone pines are canaries in a global climate coal mine. But instead of dying of carbon monoxide poisoning, as those little yellow birds once did to warn miners of deadly gases, they're growing faster—and in higher places—as a result of warming temperatures.

I saw this in living color when I hiked to the uppermost reaches of the visible treeline in the White Mountains. There, at just over 11,000 feet, I stood next to one ancient tree and scanned the area for another, then walked in the straightest route I could to reach it, looking up-mountain every few steps for signs of life.

It didn't take long to spot them, stark green against the pale gray terrain. Tiny but tough. Little bristlecones, climbing ever higher. We had changed the world, and they were chasing the thermal limits of their expanding biological niche—going where no tree had gone before, because now they could.

It can be hard to wrap one's mind around how old bristlecones are. For me, it's even harder to think about the life these little saplings could potentially have ahead of them—if we don't ruin it, of course. After all, when the ones I saw above the historic treeline in the White Mountains are as old as Methuselah, we will be nearing the year 7,000 AD.

If they make it that far, and if we are not there to stop them from doing so, they might even just keep getting older.

But not aging. That's a different thing altogether.

HOW TREES, WHALES, AND POLYPS CAN HELP US LIVE LONGER

Everything ages. We know this, right? Over time, the cells that make up every form of life on this planet begin to break down, misfire, and go just plain buggy. Eventually, they are no longer capable of sustaining life.

At least, that is how it goes most of the time. And that's how researchers assumed it went for bristlecones, too, albeit at a much slower rate. They just hadn't actually seen it. So, in the early 2000s, a research team from the US Institute of Forest Genetics began looking for signs of bristlecone senescence in every place they could think of, examining trees as young as 23 and as old as 4,713 years.

They looked in the xylem, which carries water from roots to shoots and leaves. They looked in the phloem, which moves the sugars and other metabolic products created by photosynthesis. In both they were searching for changes in the efficiency of these cellular transport tissues.

They looked for changes in shoot growth—differences in how fast and far the trees grew. They looked at the viability of pollen. They weighed the seeds. They studied how effectively the seeds germinated.

Other scientists were looking for signs of aging too. One group, from the McKnight Brain Institute, at the University of Florida, took a long, hard look at the plant's telomeres, those chromosome-protecting caps from our frog discussion, which, as they deteriorate, leave organisms, including humans, more vulnerable to age-related illnesses.[21]

They looked everywhere.

And you know what they found? Nothing. There were simply no signs of senescence.[22]

Nearly two decades have passed, and we're not a whole lot closer to understanding why this happens—or doesn't happen—than we were back then. The best guess among scientists who have studied the inner workings of bristlecones, though, is that something is happening in the bristlecones' meristems, the accumulation of cells on the ends of roots and shoots that generate new growth—the plant version of stem cells—and permit continued growth year after year.

This is what makes bristlecones so good at telling us what climates were like, long before we began keeping records. It's not just that they've been keeping track for a long time; it's that they've done it so consistently. Since bristlecones don't appear to behave differently as they age, they exist as a reliable recording device of the conditions in which they have lived from year to year, so much so that archeologists have actually used bristlecones to calibrate radiocarbon dating. The size, patterns, and density of the trees' rings, and the stable isotopes trapped within those rings, can offer us a view into past climates, water availability, humidity, and atmospheric circulation going back for thousands of years.

But organisms that don't seem to age—those that scientists say experience negligible or even no senescence—don't just tell us about our pasts. They might also be a key to our future.

To Daniel Martínez, there's no subject in science that has as much potential to be a global game changer as negligible senescence. And he wonders why more researchers aren't pursuing answers. "Unless we do not believe that negligible senescence is real," the Pomona College biologist wrote in 2012, "it seems that we should seek a better explanation for it."[23]

My friend and sometimes collaborator, David Sinclair, who studies aging at his lab at Harvard Medical School, wholeheartedly agrees. He believes life-forms like bristlecones can play a key role in helping scientists identify which human genes are the most promising targets for biomedical interventions intended to slow, stop, and even reverse the symptoms of aging. "People look at a really old tree and they think, 'Well, that's about as different from me as different gets,'" he told me in 2017. "They forget that we all emerged from the same place and, in the great scale of things, we diverged from one another on the tree of life a relatively short time ago. We carry a lot of the same genes."

Still, he said, it can be hard to get people to believe they have anything in common with organisms that, at first blush, seem so different. That's why he likes to start conversations about the comparative genomics of aging by talking about one of our relatively close cousins—the bowhead whale. "All mammals are warm-blooded, produce milk, and have a very specific brain structure that isn't found in other animals," he said. "On top of that, whales are highly social and have complex methods of communication, just like humans." Not surprisingly, we share a lot of genes—nearly 13,000 of them, including one called *FOXO3*, a variant of which has been implicated in human longevity.

While humans live longer than most mammals, bowheads blow us out of the water. With lifespans of 200 years or more, they are the longest-living mammals we know of. "What's really interesting is that the bowhead has a variation in *FOXO3* you don't see elsewhere," Sinclair said.

Once you recognize that a close cousin is doing something special with a gene we share, Sinclair said, it becomes easier to appreciate what we might learn from organisms that aren't so closely related, but also share that gene.

Among these life-forms is the creature Martínez studies in his lab at Pomona, *Hydra vulgaris,* a freshwater polyp related to jellyfish. Martínez didn't believe the rumors he first heard, in graduate school, that hydra might be immortal under the right circumstances. But since no one else was looking into the matter, he decided to disprove the notion himself.

Hydra typically grow no bigger than a half an inch, and don't last very long in the wild. So Martínez figured it wouldn't take long to prove they can't, in fact, live forever. "I thought it would take about a year and a half," he told me. "Four years later I had to publish a paper saying I was wrong."

One potential reason for the long lives of hydra? Stem cells. Hydra are almost entirely composed of them. So as long as the polyps in Martínez's lab get what they need to keep making more stem cells—clean water and a few brine shrimp to eat every other day—they can always replace old cells with new cells, and have managed to do so thus far without any sign of slowing down.

It's not enough to just have a big supply of stem cells, though. Key to Martínez's investigations of his hydras' amazing longevity is what their genomes direct those stem cells to do in response to cellular stress and when regulating the expression of genes involved in cell growth.[24] And that path of inquiry has led Martínez and other hydra researchers to *FOXO3*, which is a critical regulator of stem cells in *H. vulgaris*.

When you see one organism doing something with a gene, it might be interesting. When you see two, it could be a coincidence. When you start seeing many—and the gene they're doing it with is one we also share—that's a *lead*.

FOXO3 and its homologs in other organisms "appears especially important, forming a key gene in the insulin/insulin-like growth factor-signaling pathway, and influencing life span across diverse species," a team led by Philip Davy of the University of Hawaii's Institute for Biogenesis Research wrote in 2018.[25] When researchers like Martínez add new insights about what the gene does in organisms like hydra, the team wrote, it offers us a new way to look at "the molecular, cellular, and physiological processes that modulate aging and longevity in humans."

A few decades into his investigations, Martínez is now more convinced than ever that his initial hypothesis was dead wrong. After all, the little guys in his lab are still going strong—and other scientists are seeing similar results. In one study, Martínez teamed up with hydra researchers in Germany and Denmark to examine twelve different cohorts of hydra. Almost all of the cohorts had intriguingly low mortality rates—amounting to about one annual death in 167 individuals. Some deaths couldn't be explained, but most were the result of a lab accident, like when an individual hydra would become attached to the lid of the culture dish and dry out.

And here's the remarkable thing: That death rate didn't change regardless of whether the hydra were a year or more than forty years old. Like the bristlecone pine, they have not shown any signs of aging, even after a lot of research. One of Martínez's studies, for instance, included more than 3.9 million days of observations of individual hydra—the equivalent of looking at 100 hydra for more than 100 years.[26]

They're all still swimming along—as happy as a clam.

Well, as happy as most clams.

HOW THE DEATHS OF A FEW VERY OLD ORGANISMS EXPLAIN WHY SCIENTISTS ARE HESITANT TO STUDY SUPERLATIVES

August 7, 1964, is a day that lives in infamy among biologists, ecologists, and pretty much everyone else who has heard the story.

It was on that day that a University of North Carolina graduate student named Donald Currey broke a boring tool while trying to assess the age of a bristlecone pine that had caught his fancy near Wheeler Peak in eastern Nevada. Reasoning that the tree itself, in the middle of plenty of similar specimens, wasn't worth the chance of breaking yet another expensive borer, he asked US Forest Service officials what he should do.[27]

"Cut 'er down," he was told by a forest supervisor named Slim Hansen.

Forest service sawyers helped Currey take down the tree, and gave him a cross section of its trunk, which he took back to his motel room. That's where he started counting the rings.

Three thousand . . . four thousand . . . four thousand five hundred . . .

"And we ended around 4,900 years," Currey told NOVA in 2001, in the only interview he ever granted about what he'd done. "And you've got to think, 'I've got to have done something wrong. I better recount. I better recount again.' "[28]

He looked again and again. And slowly it hit him. He'd helped kill what was then thought to be the oldest tree in the world. The tree was at least 4,862 years old, meaning it had come into being right around the founding of Troy.

A local newspaper reporter named Darwin Lambert was furious. In an essay for *Audubon* called "Martyr for a Species," he accused Currey of murder. In the aftermath of the tree's death, Lambert later wrote, "we felt that we were walking home from a loved patriarch's funeral."

Several cuttings from the tree, also known as Prometheus and by its specimen reference number, WPN-114, wound up at the University of Arizona. Matt Salzer, the dendrochronologist who taught me about frost rings, told me there's a persistent rumor that the tree's remains are cursed. "These pieces," he said, drawing several thick sections of wood from his shelf, "came here by way of a nervous researcher."

He placed the pieces on his desk, spinning and flipping them until the ancient puzzle came together in the form of a six-foot section of what once was the world's oldest known tree. Down the center, meandering like a river, was a line marked with tiny sticky tabs, representing the centuries the tree had survived before its untimely end.

In that moment I could not have cared less if the wood actually was cursed; I had to run my fingers along the line, pausing near the year in which the United States declared its independence, and then again next to the birth of Christ. When I got to the start of the rings, I traced a circle with my index finger on wood that was 126 times older than I was, and which, but not for a rather colossal mistake, would almost certainly have outlasted me.

And I felt sad. Both for the tree and the man who had destroyed it.

Currey would go on to become a popular professor in the geography department at the University of Utah, where he was well-known for his studies of the treeless salt flats of Utah's west desert, but he carried the ignominy of what he had done to Prometheus all the way to his death in 2004—and even beyond.[29] Currey's tale has been told repeatedly over the decades. Someone even included it on a video called "5 biggest mistakes in history," right next to Mao Zedong's ecologically nightmarish Four Pests Campaign, which may have intensified China's

Great Famine, and the Union Carbide chemical spill, which may have killed as many as 16,000 people in India.

Those sorts of histrionics aside, history is always doomed to repeat itself. Sure enough, two years after Currey's death, another superlative organism died at the hands of researchers.

This time the victim wasn't an ancient tree, but a quahog clam, *Arctica islandica*, that was hauled up from the frigid ocean floor, 260 feet below the surface along with about 200 others of its kind, by researchers studying climate change. Quahogs are known to live for hundreds of years and, much like trees add rings to their trunks, the clams add a growth band to their shells for each year of their lives. Those bands carry tremendous information about the environment in which they were created; as with tree rings, the bands are bigger when growth conditions are more favorable.[30]

Quahogs are also among the most commonly fished clams (if you've eaten clam chowder, you've likely digested the flesh of an animal that was hundreds of years old before it was caught), so the scientists didn't think it was a big deal to immediately throw all of their samples into the freezer on their boat, just like fishermen do.

It was only when they got back to the lab and started to count the bands that they realized that among their catch was a clam older than anything they'd ever studied before.

An initial count put the clam's age at 405. A second look, plus radiocarbon dating, added another century. The clam, dubbed "Ming"—for it was during that Chinese dynasty that it had been born—was 507 years old when it was killed.

The researchers who killed Ming have so far been spared the level of notoriety bestowed upon the man who killed Prometheus, but they've had their fair share of haters, too. *The Independent* called Ming's death

"A Clamity!" Others were less kind. "We've had emails accusing us of being clam murderers," marine geologist James Scourse told the BBC.[31]

If scientists didn't already have plenty of reasons to be reluctant to search for and study superlative species, the tales of Prometheus and Ming might offer even more pause. Research always carries a risk to the thing being studied; anyone who tells you otherwise is prevaricating. And when scientists screw up something superlative, whether that screw-up is the result of actual negligence or happenstance, the reaction is going to be amplified

And yes, the felling of Prometheus and the freezing of Ming were accidents, and might have been avoided—in both cases the research objectives likely could have been achieved without killing the things being studied. But both instances came with significant scientific benefit.

The sections of Prometheus I saw in Salzer's lab have been used to build dendrochronologies that are helping us understand past, present, and future climates. Ming's shell has been used in the same way—by comparing the patterns of its bands to those of other quahogs, scientists were able to show that human-caused climate change has begun to disconnect marine and atmospheric systems that, before we started mucking things up, always operated in sync.[32] That's an exceptionally important finding, one that speaks to a tragedy far greater than the death of a very old animal.

And it's worth noting that Ming actually wasn't the oldest animal ever found. Not even close. That record belongs to another ancient inhabitant of the sea.

HOW THE WORLD'S OLDEST KNOWN ANIMAL IS HELPING US UNLOCK THE OCEAN'S DEEPEST SECRETS

The oldest animal in the world doesn't look much like the things most people see when they picture animals. It doesn't have a mouth or eyes. It doesn't have legs or flippers.

But *Monorhaphis chuni* is an animal. And it's really old.

M. chuni is a member of a class of animals called hexactinellids, also known as glass sponges. I suppose its top is fairly sponge-like. To me it looks like a big tan loofah. But the bottom resembles something Superman's ancestors on Krypton might have taken to battle, sort of like a 9-foot glass throwing spear. That's the sponge's silica spicule, a long, skeletal leg that attaches it to the bottom of the ocean.

This sponge is at the heart of yet another story about the accidental killing of an ancient thing. It had been living a peaceful and very long life at a depth of more than 3,500 feet in the Okinawa Trough in the East China Sea when, in 1986, it was unceremoniously dredged up. At the time it was provided to the Chinese Academy of Sciences, no one knew what to do with it. It was regarded as an oddity—the longest hexactinellid anyone had ever seen. For that reason, it was fun to take photos with. That was about it. The sponge spent the next quarter century on a shelf.

Some researchers had hypothesized that hexactinellids could potentially reach ages upwards of 20,000 years, which would, if confirmed, put them squarely on the throne of the world's oldest animal. But until rather recently no one knew how to test that theory, and since hexactinellids are rather hard to get a hold of, no one had put much thought into it. A few years ago, though, a paleoclimatologist named Klaus Jochum, who had been on the hunt for novel ways of understanding ancient climates, heard the academy had possession of the longest intact hexactinellid spicule anyone had ever seen. So he asked to take a look.

Cross sections of the cylindrical silica leg revealed a concentric growth pattern, just like a tree's. And the rings were different sizes and widths, just like a tree's. But even under intense magnification it was hard to see where one silica layer stopped and another began, and without a lot of other samples of *M. chuni*—preferably ones measured over time in a place where the seafloor climate was being closely monitored—it wasn't clear whether the rings appeared annually, as they do on a tree and as bands do on a clam, or at some other rate of growth.

But when Jochum's team members tested spots along the various rings for oxygen isotopes and magnesium-to-calcium ratios—common proxies for ancient sea temperatures—they saw something fascinating. At the youngest, outer layers of the spicule, their analysis offered an inferred temperature of 4 degrees Celsius, which aligned quite precisely with the environment at the bottom of the Okinawa Trough at the time the specimen was taken from the deep. As they tested the rings closer to the center, they saw four spikes in inferred temperatures, likely the results of temporary hydrothermal activity. Overall, though, the scientists saw a very gradual shift remarkably consistent with other research conclusions about how the seawater in that region of the world has been slowly warming since the last ice age.

And, at the very center, where the skeleton is oldest, the tests suggested a temperature of 1.9 degrees Celsius—which scientists believe to have been the temperature in that part of the deep sea 11,000 years ago. This long-dead sponge was the world's oldest thermometer.[33]

As scientists should be, they were careful with their proclamations, offering a 3,000-year window on either side of their estimation. The specimen could have been as old as 14,000 years and as young as 8,000. Even taking the most conservative estimate, though, it was the oldest animal ever identified, and by a longshot. Fifteen Mings could have come and gone in its lifetime.

When this sponge was born, humans were a distinctively social, problem-solving, tool-using hominid, but we were also a species that had yet to make a significant impact on the planet itself. By the time the sponge died, our species was deep into a 200-year reign of terror that has led to a global mass extinction and a rapid shift in climate.

That makes *M. chuni* an exceptionally valuable resource. Just as bristlecones can offer us a view of how we have changed the climate 11,000 feet above sea level, and in the same way that quahog clams can tell us what our impact has been in the shallow sea, sponges offer us the potential to understand the paleoclimate of the deep ocean—and stand as witnesses to our impact there, as well.

That's not all they can teach us.

WHAT SPONGES, TREES, AND WHALES CAN TEACH US ABOUT HUMAN LONGEVITY

At the heart of *M. chuni*'s remarkable success, individually and as a species, are three things: simplicity, stress, and cellular survivability.

The specimen that was dredged up from the Okinawa Trough lived a really simple life, its movements confined to the slow currents of the deep sea. Hexactinellids don't even have the tiny, ever-spinning flagella that other sponges use to pump water and nutrients through their bodies—the glass sponges simply accept whatever tiny particles of food the ocean offers, which would have amounted to a similarly tiny meal, every day, for the roughly 4 million days of the sponge's life.

A simple life is not necessarily one that lacks stress, however. Indeed, there is perhaps no environment in the world that exerts greater physical pressure on an organism than the deep sea. At the bottom of the Okinawa Trough, the pressure is nearly 1,500 pounds per square inch, and the temperatures may have fluctuated, over the millennia of the sponge's long life, from as low as a darn-near-freezing 0.8 degrees Celsius to as high as take-a-swim-to-cool-off 10 degrees Celsius. But like a boxer getting ready for a bout—training every day for hundreds of millions of years—hexactinellids have been "working out" against these stresses for a very long time, resulting in organisms that are as tough as a heavyweight champion. A less stressful environment would have produced a less fit species, one that simply couldn't survive for so long.

Stress can be the very thing a species needs to evolve into a long-lived being—provided, that is, that there's an easy way to regenerate when their cells succumb to that stress. *M. chuni* has that going for it, too. Sponges are packed with stem cells. So sure, they might exist in an environment akin to living next to a madly swinging wrecking ball, but they also have a built-in brick factory with which to rebuild.[34]

The simplicity-stress-survivability equation doesn't just apply to sponges. We see it in all of the other long-lived organisms we've been discussing, too. Aspen, for instance, live pretty simple lives, starting with their very genes. *P. tremuloides* has one of the shortest genomes among trees, with just 550 million base pairs. And, as you might recall, neither the aspen nor two other massive and ancient species, *L. tasmanica* and *G. renwickiana*, let life get complicated by trivialities like sex. They've figured out a simpler way to survive, and thrive.

If you were to spend a night in Pando's enormous embrace, though, you'd be treated to a visceral understanding that its simple life isn't a stress-free one. It gets really cold at 9,000 feet above sea level, and the forest floor disappears for months at a time each year under a thick blanket of snow. And even before we killed off the wolves and cougars, there were always deer and elk there that liked to munch on aspen shoots. Add to that the fires that have swept through the forest for millennia. That's some serious stress.

But Pando has a steady supply of the kingdom Plantae's equivalent to stem cells—meristematic cells, which are found at the tips of roots and shoots all over the organism. Cut down a stem, burn it to the ground, eat through it with beetles, chomp it up in the mouth of an ungulate—whatever—and these undifferentiated cellular supply points will get immediately to work, dividing rapidly to produce new shoots. University of Barcelona biologist Sergi Munné-Bosch, an expert in plant senescence, once described meristems as "the kings" in a botanical game of chess—so long as one meristematic cell remains alive, the game continues. All other tissues, he wrote, "will play an altruistic role to serve the meristems."[35]

The bowhead whale is yet another example of the longevity trinity at work. It too lives a relatively simple life. Unlike most other whales, it doesn't migrate; it lives its entire life in arctic and subarctic seas. It is among the slowest swimming and least social cetaceans. It also lives a life of constant stress. It's damn cold in the Arctic Ocean, and zooplankton can be hard to come by for long stretches of the winter. And, sure enough, when an international team of geneticists sequenced the bowhead genome in 2015, its members found species-specific mutations that appear to promote DNA repair, cell-cycle regulation, cancer suppression, and aging—a well-stocked armory of genetic weapons with which to do battle against the wear-and-tear of a long life in the frigid sea.[36]

Again and again, research has shown that organisms that manage to eke out longer existences in this world benefit from a combination of very basic lives, challenging environments, and cells that can "turn-to" when replenishment is needed.[37]

Can humans similarly master the simplicity-stress-survivability equation? John Day thinks so. And when the Stanford- and Johns Hopkins–educated cardiologist met me in southern China in 2016, he was determined to prove it.

He started by introducing me to Matao, who was spiritedly walking back and forth between her home and a spot by the river where she had been preparing vegetables by the armful: squatting down, cutting the greens, scooping them up, and then going back for another load—and another, and another. And smiling all the while.

I might have pegged her for a spritely eighty. In fact, Day told me, she was 101.

I couldn't help but laugh. "Chuckle all you want," Day said. "She's the *youngest* of this village's centenarians. And not even the most active."

The elders of Bapan, which straddles the tranquil Panyang River near the Chinese border with Vietnam, don't have formal records of their births. Owing to this, Bapan was passed over as a "Blue Zone," a term coined by demographer Michael Poulain to refer to places around the world where people live abnormally long lives. These are places like Okinawa, which has approximately one centenarian for every 2,000 residents. By Day's estimate, though, Bapan might be one of the *bluest* places in the world. It is a town where about one in every 100 residents has reached the century mark, and plenty more are right in line behind them, living incredibly healthy and active lives well past the point at which many people in the Western world would tell you they'd *rather* be dead.

Among the villagers were several who had surpassed 110 years.[38] The eldest at the time of my visit, named Boxin, was reportedly 116 years old, and still waking every day from his wooden bed mat to greet the travelers from across China who make pilgrimages to this place to learn the secrets of his remarkably long and healthy life. But what Boxin tells these travelers is that there really *aren't* any secrets, just some really good lessons for life.

Day has distilled these lessons down to seven basic principles built around food, motion, mindset, community, rhythm, environment, and purpose. "Underlying all of these things is simplicity," Day told me one day as we walked across a gently swaying footbridge to meet with some farmers on the other side of the river. "The people here don't need exercise regimens and dieticians to help them live healthy lives; they *simply* live healthy lives."

That doesn't mean they've had stress-free lives. Much to the contrary. The elders here work seven days a week in their fields, and do so well into their nineties and hundreds. Over the decades, they have faced war and political persecution. During the Cultural Revolution, some were tortured and others threatened with execution.

Simplicity? Check.

Stress? Check.

All that was missing from the universal longevity equation, as I was coming to understand it, was some manner of superior cellular survivability.

"Funny thing about that," Day told me. "The tests I've run, the studies I've been reading, they don't indicate *anything* different about these folks. They're no more genetically equipped for longevity than you or me."

That doesn't mean their bodies aren't unusually well-equipped for cellular survivability. It means *all* of our bodies are—or can be.

When we eat fresh and unprocessed foods. When we live lives of constant motion. When we approach the world with optimism, surround ourselves with people we love, live our lives in a reliable rhythm, seek out healthy environments, and find purpose in our lives. When we do these things, our cells become survivors. And that's not just a gift we give ourselves. Both socially and epigenetically, through the power of inheritable genetic expression, we're handing good habits and healthy genomes to our children, grandchildren, and great-grandchildren—and increasing the odds that we'll get to spend a lot more years with them.

Chapter IV

FAST TIMES

Why the Quickest Animals Probably Aren't the Ones You Think

The cheetahs at the Ensessakotteh wildlife refuge went through hell before arriving at the conservation and education center run by the Born Free Foundation in central Ethiopia.

Poachers had killed their mothers. Smugglers had stuffed them into tiny crates, wicker baskets, and buckets without air holes, apparently intending to transport them to the Middle East, where they have long been prized as pets. Ethiopian authorities ultimately rescued the cubs, but not before many of their siblings had died.

When I learned their story, it seemed to me that the cats I was about to meet would be shells of the ones they could have been had they not been stolen from the wild. I was joyfully mistaken. The Ensessakotteh cheetahs may have been robbed of the maternal nurturing and natural conditioning they needed to fend for themselves in the bush, but they slept together in the sun, frolicked in the tall grasses between the trees, and took prowling crouches when wild crows and hawks dared to land nearby.

And as the sun began to dip and it became time to feed, they did what cheetahs are born to do.

There is no way to adequately describe the power of a cheetah's first steps into a sprint. It is as though their entire bodies have been drawn back, like an arrow on a bow, and then released into the wind. Nothing so explosive should be so quiet, and yet they run with barely a sound, just the swishing of grass against their backs and the scratching of dirt beneath their feet.

Every schoolchild can tell you that cheetahs are the world's fastest animal. But fastest is a quirky concept. Quirkier even, perhaps, than biggest, smallest, or oldest—for there are hundreds of ways to think about speed.

For years we called Usain Bolt "the world's fastest human" because he was nearly untouchable in the 100-meter dash, but if you had put him in a mile race with Taoufik Makhloufi, or a marathon with Eliud Kipchoge, or an ultramarathon with Kathleen Cusick, Bolt would have been left in the dust.

Put any of those people in a pool with swimmer Katie Ledecky. Or put Ledecky in open water, next to long-distance phenomenon Chloe McCardel. Or put McCardel in an airplane, and let her dive out the door alongside record-holding speed skydiver Henrik Raimer. You'd have a different "fastest" every time.

There is no world's fastest human, because there are different ways to measure speed in humans. And there are even more ways to measure speed in the rest of the natural world.

Cheetahs are remarkable beasts. They can exceed 60 miles per hour in very short bursts. (The land speed record for a 100-meter dash was set by a cat named Sarah in 2012; she reached a top speed of 61 miles per hour in that sprint.) When researchers collared five wild cheetahs, though, and tracked their running habits, they learned that the cats usually run at only about half that speed, and can only do that for a few hundred meters. Then they're done, often for the rest of the day.[1]

So, no, *Acinonyx jubatus* isn't really the world's fastest animal. In most ways of thinking about speed—the rate at which an object covers distance—it's not even close.

But to see a cheetah run is to stop caring about any of that. Because to see a cheetah run is to witness perfection of design, from nose to tail.

Let's start up front. Most cats have a relatively small naval cavity, consistent with predators that prioritize other senses over smell. In the cheetah skull, though, the nasal cavity is a gaping hole. In this way, a cheetah is a lot like a fighter plane.

When the Navy wanted to make its workhorse jet, the F/A-18 Hornet, even faster, one of the first things it did was make the aircraft's intake ramps bigger. This allowed for more air to flow into the airplane's compressors, which increase the pressure of air as it moves through the engine's combustor. More airflow means more thrust. The Super Hornet was born.

Animals work in a similar way. Air, after all, carries oxygen, the life-sustaining gas we need, and must constantly replenish, in order to function.

Of course, it doesn't do much good to bring in more air if there's nothing that can be done with it. So a cheetah's thoracic cavity is filled to the brim with big lungs, a hefty heart, and a large liver with which to take in, move, and use that oxygen to mobilize glycogen, which provides big bursts of energy.[2]

To keep that intake-to-engine passageway straight, and maximize efficiency, cheetahs keep their heads almost perfectly still when they are running—despite the fact that almost everything behind the cheetah's head is moving like crazy, thanks to a spring-like spine.

Cats in general have exceptionally flexible vertebral columns; even if you don't own a house cat, you can probably picture the classic Halloween decoration of a cat arching its back, a common feline reaction to fear. Most cats use that flexibility to stretch out their muscles—something they need to do after sleeping for three-quarters of each day or longer. Cheetahs, though, use that flexibility to stretch out their legs when they run.

To get an idea of what this looks like, cup your hand downward as if picking up a baseball, then raise your fingers up as far as you can. That's how a cheetah's spine flexes when its front and back legs cross, then spread apart, as it runs in a rotary gallop, a four-beat gait common to mammals. Like a spring being sprung, cheetahs shoot their legs outward, extending their spine so far that it doesn't just straighten, but slightly reverses its curve. This allows cheetahs to extend the length of their strides. Cheetahs can cover more than 20 feet with each stride. By way of contrast, a long running stride for a human is about eight feet.

The length of those strides doesn't come just from the spine, though. Cheetah legs include a unique one-two punch of muscle fibers that substantially differ between the front and back limbs. The back limbs have predominantly fast-twitch fibers, which can create tremendous force but have low endurance, while the front limbs have a larger share of slow-twitch fibers, which offer little force output, but are resistant to fatigue. But the cheetah's front *paws* are more like its back limbs: they have a much larger supply of fast-twitch muscles than the rest of the front limbs do—allowing the cat to stay balanced at high speeds.[3] Essentially, researchers say, cheetahs are like rear-wheel-drive cars with high-performance steering.[4] And with good tires, too—at the ends of those legs are hard pads and claws that don't retract, helping cheetahs turn at exceptionally high speeds.

And finally, there's the tail, which might be the most amazing, and most underappreciated, aspect of the cheetah's ability to run down its prey. When *A. jubatus* runs straight, its tail stays directly behind it. But when it shifts direction, even a bit, it uses the rear appendage as a whip-like counterbalancing tool. The tail snaps left, and the cheetah goes right. The tail snaps right, and the cheetah goes left. The tail snaps again, and a gazelle becomes dinner.

When it comes to speed, though, there might be something even more important than a cheetah's design: its size.

Notwithstanding the seeming flawlessness of the cheetah's design as a creature built for speed, at first the world's fastest cat presented a problem for biologists trying to understand the role of speed in the animal kingdom: Why doesn't absolute speed grow as animals increase in body size? If a house cat can run at 30 miles an hour and the larger lynx can run at 50 miles an hour and the larger-than-that cheetah can reach 60 miles an hour, why can't the even-larger-than-that tiger run even faster?

Ecologist Myriam Hirt believes larger cats *could* be faster—theoretically, at least. Large animals generally have more fast-twitch muscles than their smaller peers, and would be able to use those muscles to accelerate for longer periods, if only the oxygen that supplied those muscles didn't run out so quickly.[5] If a tiger could dope its fast-twitch muscles with a rapid resupply of oxygen, it would likely be the fastest cat in the world by a factor that correlates to its massive body size. Alas, Hirt and her colleagues from Germany's EcoNetLab have theorized, in the real world, the fuel needed to move a big body runs out long before a maximum theoretical speed can be reached.[6] Their theory suggests that there's a "sweet spot" between being big enough to take large and powerful strides and small enough to convert oxygen efficiently to muscle motion.

The theory doesn't just work for cats. And it doesn't just work for mammals. In fact, a focus on mammals alone may have been why this relatively simple insight wasn't reached much earlier—the correlations are a lot less obvious when you're looking at animals with a limited size range and a lot of other physiological similarities.[7] So, before publishing their findings, Hirt and her fellow researchers applied their calculations to nearly 460 running animals of all sorts, including birds, arthropods, reptiles, and mammals. Plotting these creatures' maximum speeds against their body masses resulted in an inverted J-curve that starts with tiny insects and moves upward with sublime consistency until peaking at about 60 miles per hour and 100 pounds—that's the cheetah, of course. The curve dives from there toward larger and slower animals, like moose and hippos and elephants.

But Hirt didn't stop with the runners. She and her team created similar plots for swimmers and flyers, again without respect to what part of the animal kingdom those creatures came from. The swimmers included birds, reptiles, mammals, arthropods, fish, and mollusks. The flyers comprised birds, arthropods, and mammals. Although the peak in speed and size was a bit different for those groups, a very similar curve appeared. In each set of animals, the as-it-gets-bigger-it-gets-faster plotline held strong until peaking about three-quarters of the way through the group, where it began to peter out.[8]

Because it holds up among animals we know, this model was quickly applied to animals we don't know as well, including ones we've never actually seen. And that, I'm afraid, puts yet another blotch on the already-well-blemished science behind one of my family's favorite movies, *Jurassic Park*.[9]

"Well," the character John Hammond brags early in the first film of that fantastic franchise, "we clocked the *T. rex* at 32 miles per hour."

That purported speed is put on display in one of the film's most famous scenes, the reason I chuckle any time I notice the words "Objects in mirror are closer than they appear" on the side-view mirror on my car.

But if Hirt's theory holds for carnivorous theropods, tyrannosaurs aren't likely to have been able to break 19 miles per hour.[10] Of course, that still makes them faster than most humans. So *T. rex* would still likely win a foot race with *H. sapiens*; it would just be a bit of a longer run before lunch.

Dinosaurs aren't the only creatures about which we have a better understanding thanks to cheetahs, though. A lot of what we've learned about how animals survive over evolutionary time comes from our studies of *A. jubatus*.

WHY CHEETAHS SHOULD BE EXTINCT, BUT AREN'T

The Late Pleistocene extinction event was tough on the creatures of our world. By some estimates, about three-quarters of large mammals—those about 90 pounds or bigger—were lost in the Americas and Australia. In Europe and Asia, the toll was closer to half.

African animals were better at weathering the storm. Only about a sixth of that continent's large mammals died out, but many of the rest struggled to survive. Cheetahs, in particular, were on the ropes. Their population fell so low that widespread inbreeding was the only way to survive.

Such a survival strategy, of course, has diminishing returns—as demonstrated by the last mammoths on Earth, those that survived the

Pleistocene catastrophe and continued living on Wrangel Island, in the Arctic Ocean, until about 4,000 years ago. Scientists believe those woolly beasts might still be with us today if their genetic diversity hadn't been so depleted, resulting in "genomic meltdown."[11]

It's not clear whether cheetahs were still in the process of accumulating more detrimental mutations, or if they were instead in the midst of the glacially slow process of recovery, when the next big extinction event (that would be us) came along. What we do know is that the result of the Pleistocene population bottleneck is a species suffering from extreme genomic depletion in just about every measurable way, from a dearth of single-nucleotide variants, to a lack of diversity in mitochondrial DNA, to a paucity of cell surface proteins that support their immune responses. If there's a silver lining to that latter category of genetic monotony, which is known as histocompatibility, it's that cheetahs are very good at accepting skin grafts from other cheetahs—almost as though the cats are all siblings.[12]

And genetically speaking, they pretty much are.

When geneticist Stephen O'Brien first studied the genomes of dozens of cheetahs in the 1980s, he was confused by what he saw. "You guys did not really collect fifty cheetahs, did you?" he later recalled joking with Mitch Bush, the head veterinarian at the National Zoo in Washington, DC, who had coordinated the samples for the study. "What you actually did was to collect one cheetah's blood then split the blood into fifty separate tubes, right?"

The cats were almost genetically identical, O'Brien wrote in his book, *Tears of the Cheetah*. "Their genes had the look of deliberately inbred laboratory strains of mice or rats."

The genome sequences of wild-born cheetahs are, on average, 95 percent homozygous, perhaps the least diverse mammal genome in the natural world. By comparison, the critically endangered Virunga

Mountain Gorilla is 78 percent homozygous, and the heavily inbred Abyssinian cat is 63 percent homozygous.[13]

The result of all of this genetic similarity is a population of animals with exceptionally high cub mortality rates, and with far more disease susceptibility than their fellow cats.[14] And that, of course, was the situation facing cheetahs even *before* the human population explosion in Africa began to take its toll.

In the early 1900s, there were about 100,000 wild cheetahs stretched across Africa and Asia. Today there are about 7,000 left. And only two population groups offer any real hope of not inbreeding themselves out of existence[15]—one in southern Africa that has about 4,000 members, another in the Serengeti that has about 1,000. The other African clusters are small and shrinking, and there is likely no functional population left in Asia.

As you might suspect, poachers and traders are a big factor in this population freefall, but so are farmers, who kill the cheetahs to protect their livestock; when it comes to predator controversies, cheetahs are to Africa as wolves are to North America. Of nine free-roaming, collared cheetahs that were followed for a study published in 2017, four were shot by landowners.[16] Roads are another huge hazard. When researchers tracked a population of cheetahs for two years in 2011 and 2012, they found that more than a quarter of verified deaths came as a result of being hit by cars or trucks.[17]

Put all of this together, and the result is a current rate of population decline some researchers believe to be 10 percent per year. If cheetah numbers continue to decline at that rate, half the world's population could be gone in the next decade.

It is for these reasons that Zelealem Tefera Ashenafi, Born Free's chief representative in Ethiopia, is circumspect about the chances of success for a release of the foundation's rescued and rehabilitated cheetahs.

"Teaching them to hunt is a challenge, of course, but it is a challenge I believe we can overcome," he told me. "The bigger problem is: Where do we take them? Where can we bring them where they will be safe, even relatively safe, from the threats that are destroying the cheetah populations everywhere?"

Still, he said, he wants to try. "Because," he said, "what good is it to keep these cheetahs in our conservation center if they don't actually help us contribute to conservation?"

He has been eyeing Awash National Park, about 100 miles to the east of Ensessakotteh, as a potential reintroduction area. "It's close enough that we can vigilantly protect the cheetahs," he said. "We can keep monitors there and give the cheetahs a soft release, with radio collars and food if, it turns out, they cannot provide for themselves."

Still, Ashenafi lamented, the odds do not seem to be in the cheetah's favor.

Yet in a century in which so many of their fellow vertebrates went extinct, and even as their very genomes conspired against them, cheetahs have managed to survive.

How? The answer may lie in the same lack of genetic diversity that also threatens their survival. That shallow selection of DNA may have essentially locked in the cats' "speed genes," which code for adaptations in muscle contraction, stress, and cardiopulmonary responses. There simply aren't any "genetically slow" cheetahs who, when bred with the fast ones, would bring down the pack. Maintaining that speed over evolutionary time, even though they might not have needed it to catch much-slower prey, could have given cheetahs an evolutionary über-advantage that counteracted their genetic disadvantages.

"Some people say, 'Well, now that this bottleneck has happened the cheetahs are totally screwed,' " O'Brien told me in 2017 while he was back in the United States during a break from his duties at St. Petersburg State University, where he is working to get more Russian scientists engaged in genomic analysis projects. "But I say that's not necessarily true."

A population bottleneck, O'Brien said, "is a little like a poker hand. Most of the time your cards are going to be mediocre, and sometimes you're going to get a complete bust. But every bottleneck is a deck reshuffle, and sometimes you draw a card that makes your hand."

Cheetahs were dealt an unusually good hand. "They retained a hell of a lot of good genes," he said.

What we continue to learn, as we work to understand how cheetahs have managed to survive a shallow genetic pool, will undoubtedly help us address the challenges faced by other animals whose genetic diversity will suffer as their numbers shrink.

Of course, what we can learn will be badly diminished if cheetahs go extinct.

So the race is on.

WHY PRONGHORNS ARE ALWAYS RUNNING FROM GHOSTS

The pronghorn didn't approach so much as appear before me. And in playing that moment back in my head, I still cannot figure out how we came to be standing face-to-face and just feet away from one another on that dusty bluff.

We stood there looking at one another for a moment. I tilted my head to the right. He dipped his down, then lifted it again and leaned back, exposing his muscular shoulders and the stark white triangle of fur on his chest. His coffee-brown horns were as long as the distance from my elbows to my fingertips, and rounded toward one another, nearly touching at the tips.

He was magnificent. The biggest pronghorn I'd ever seen—and I've seen quite a few. It's hard to miss them in the Red Desert in southern Wyoming, the home of the largest migratory herd in North America.

I'm not sure how long it lasted. A minute perhaps, though it could have been five or ten. We took one another in. Squinting at each other in the late afternoon sun. For me, the whole world slowed down, and I wondered if that was his experience as well.

And then it happened. Something stirred in the sagebrush behind him—a jackrabbit maybe—and the pronghorn took off, bounding over scrub and rocks, left then right then hard left along a wash, leaning into the turns. In a matter of five or six seconds he was gone.

Pronghorns are known to reach a top speed of 55 miles per hour. That's not quite as fast as a cheetah, but it's not far off either. Pronghorns, though, can keep up the speeds that cheetahs *typically* reach for a lot longer. Cheetahs can run at 40 miles an hour for a few hundred meters; pronghorns can keep up those sorts of speeds for miles.[18]

How? Before answering that question, we need to have a talk about taxonomy.

Like many schoolchildren, my earliest mental images of the American West were largely informed by Brewster Higley's poem "My Western Home," which later became the song "Home on the Range." But it turns out that Higley's most famous lyric—"give me a home where the buffalo roam, where the deer and the antelope play"—did a bit of a disservice to our collective understanding of what animals actually live in the Great Plains. For there are no buffalo roaming the range in the United States, and there never have been. Like the African cape buffalo and the Asian water buffalo, the American bison is a member of the Bovidae family. But so are gazelles, sheep, and antelope—the latter of which is also not native to the United States. The "antelopes" we see on the plains are properly pronghorns.[19]

Culturally, these are semantic distinctions. There's no good reason to stop referring to the five-cent pieces minted from 1913 to 1938 as "buffalo nickels." And we don't have to stop calling the largest island in the Great Salt Lake—where pioneer hunters reportedly found pronghorns so plentiful that it was hard to miss them—Antelope Island.

Scientifically, though, it's important to understand that similarities in appearance between one life-form and another aren't always reflective of a closely shared evolutionary history. So our biologist predecessors did us no favors when they, on occasion, not only made bad relational assumptions but enshrined those assumptions in taxonomic nomenclature, as they did in the case of the pronghorn, *Antilocapra americana* of the family Antilocapridae.

We live in a world in which comparative genomics is offering deep insights into the codes upon which our very existences are based. But hunting for shared DNA in two genomes is like searching for a few specific lines of text in an enormous library of books. And in such a search, it doesn't help anyone if some of the books are mislabeled or have been put on the wrong shelves.[20]

So we should be wary of assumptions made by our forebears, especially those that take on cultural gravitas, which can cloud our scientific perspective for generations. That was the case, as we know, with giraffes—which, it so happens, are far more closely related to the pronghorn than the antelope is.

To avoid predation, giraffes didn't just evolve to be tall, but also to be very strong kickers—they can kill a lion with a single strike—and very fast runners. And pronghorns evolved the ability to run at high speeds for the same reason.

But if you only look at the predators that pronghorns have to contend with on the American prairie, you might wonder why they need to run so fast. Wolves and coyotes are quick, but they're not nearly as fast as pronghorns. Even juvenile pronghorns can often outrun those predators.

So why do pronghorns need to drive 55?

The answers, some researchers believe, are *Miracinonyx inexpectatus* and *Miracinonyx trumani*, big cats belonging to a genus first identified by Edward Cope—of Cope's Rule fame—that evolved alongside pronghorns for millions of years in North America, and that are sometimes called "false cheetahs."[21]

The *Miracinonyx* cats, which were cougar-sized but cheetah-shaped and thus assumed to be really fast, died out about 11,000 years ago. But, the way zoologist John Byers sees it, their legacy lives on in the DNA of one of today's best runners.

Pronghorns, Byers notes, "are truly Olympian runners in a world of notably less-than-Olympian predators."[22] And the reason they're so overbuilt, he believes, is because they evolved as runners during a time in which North America was filled with predators like *Miracinonyx*.[23]

Just about every animal is carrying some sort of evolutionary baggage it just doesn't need anymore. Humans have tailbones, grow wisdom teeth, and get goose bumps, none of which are good for much these days. Given enough time, it's reasonable to assume that we'd shake off these traits.

But traits born of a need for survival—not those that make life easier or a bit less dangerous, but those that acutely prevent death and extinction—seem to be a lot harder to get rid of when they're not needed, or not needed as frequently. That might explain why, even though our day-to-day lives aren't usually interrupted by attacks from saber-toothed cats or giant primate-eating eagles,[24] we can still rely on our sympathetic nervous system to give us a mega-dose of adrenaline and norepinephrine for surviving our much-less-common life-and-death situations.

And it means we get to see pronghorns run. Not just on the range, but on treadmills.

Yeah, treadmills.

That's all thanks to a guy named Stan Lindstedt. With the evolutionary trigger for the pronghorn's need for speed seemingly identified, the Northern Arizona University physiologist wanted to understand the biomechanics that allowed that evolution to happen.

Because pronghorns don't really look like great runners. They look like furry sausages with spindly legs. Pronghorns aren't shaped all that differently from goats, and no one would expect a goat to enjoy running.[25]

Just to be sure, though, Lindstedt checked. Sure enough, goats don't like treadmills. They'd only run when the researchers bribed them with a lot of food.

But when Lindstedt and his team put the pronghorns on the treadmills, they not only ran, but seemed to love doing so. "You'd open the door to the lab," Lindstedt once told the *New York Times*, "and they'd run right in and jump on the treadmill."[26]

Unlike goats, Lindstedt found, pronghorns are perfectly built machines for transporting oxygen. Pronghorns have bigger tracheas with which to draw in oxygen. They have bigger lungs with which to absorb that oxygen. They have more hemoglobin with which to transport it to their muscles. And the cells in their muscles have a denser concentration of mitochondria with which to fuel contraction. Sure, pronghorns don't look like speed machines, but looks can be deceiving.

Just ask the lowly mite.

WHY THE WORLD'S FASTEST RUNNER IS LIKE BATMAN

Anyone who has seen a cockroach scurry across the floor when the basement lights come on can attest to how fast these critters are. Although they have wings and can fly, they really don't need to. They're far slower in the air than they are on the ground.

For a long time, Guinness offered the American cockroach, *Periplaneta americana*,[27] the title of "fastest running insect." Over time, though, researchers began to realize that—when it comes to hunting, and surviving the things that are hunting you—an animal's speed relative to its body size is almost always more important than absolute speed. To determine a true fastest insect, an entomologist named Thomas Merrit gathered data from fellow insect researchers and crunched the numbers relative to each bug's length. The resulting figures showed *P. americana* could move at a rate of 50 body lengths per second.

And that was fast. By comparison, after all, cheetahs can only run at about 16 body lengths per second. But in another insect, Merrit found a new champion. The Australian tiger beetle, *Cicindela eburneola*, could run at more than 170 body lengths per second. In relative terms, a six-foot man would have to run 720 miles per hour to be as fast.[28]

But records are made to be broken. And nowhere is this more true than in nature, where wonders truly never cease.

In Claremont, California—not far from the lab where Daniel Martínez is maintaining a nearly immortal collection of hydra—a group of researchers led by an undergraduate named Samuel Rubin took note of a tiny mite, *Paratarsotomus macropalpis*, dashing across the sidewalks in the SoCal heat.

The mite was no secret. It had first been identified in 1916, and lived in the middle of one of the most highly populated metropolitan areas in the world. But no one had previously paid any scientific attention to it.

A search of the ScienceDirect database through the early 2010s offers zero articles that so much as mentioned the little guy.

"They're pretty easy to overlook," said Jonathan Wright, Rubin's former adviser at Pomona College and his co-author on the paper. "They're tiny, about a millimeter long, and they move so quickly when they are running that, if you were not scrutinizing them carefully, you might just conclude they were a bit of blowing dust."

They're also darn hard to catch. Wright told me he's thankful the mites like sidewalks and driveways, because trying to capture them in their natural sandy habitat generally results in in an aspirator jar full of dirt—and often no mites.

The team was able to get some video, though, of the mite darting about on the sidewalk. And when the researchers used that footage to trace a path behind the animal, then measured the distance it had traveled, they were floored.

At first Rubin thought he'd miscalculated. But when he ran the numbers again, it was clear: He'd helped discover the world's fastest runner.

The mite was moving at up to 322 body lengths per second. The equivalent human speed would be something in the neighborhood of 1,300 miles per hour.[29] To put this into a bit of context: There are only about a dozen reported times in history that an *airplane* has gone faster than 1,300 miles per hour.

P. macropalpis wasn't just leaping from one place to another. It was legitimately running—with a stride rate of 135 steps per second, the highest cycling of any weight-bearing muscle ever reported in any animal.[30] By way of comparison, human sprinters take about three steps per second. Even the famous "Jesus lizard," the basilisk, which moves so fast it can walk on water, only takes about twenty steps per second.

By reinforcing the scientific theory of scaling—which suggests that the smaller organisms get, the less force they need to increase speed—the discovery of the mite's amazing stride was soon being used to inform our understanding of the potential of nanomotors, organic engines capable of converting energy to movement at a molecular scale.[31]

P. macropalpis might also help teach us to create machines better adapted for acceleration, deceleration, and rapid turning. The tiny mite, after all, can stop in an instant, and can turn at speeds and angles that would rip other animals' legs off.

To understand how it does that, Rubin and Wright slowed down the footage and blew it up big. When they did, they saw the mite had two strategies for turning, one for fast turns and one for really-stinking-fast turns.

The fast strategy is like an exaggerated version of how marching bands turn. To pivot in a column, the musicians on the inner part of the turn shorten their gait to the point of nearly marching in place, while those on the outside must lengthen their stride. The mite was doing the same thing—taking very short steps with the legs on the inside of the turn and longer steps with the legs on the outside.[32]

The really-stinking-fast strategy is straight out of Tim Burton's now-classic 1989 reboot of Batman, when the caped crusader deploys a grappling hook to help the Batmobile make a sharp turn during a car chase. Instead of a hook and rope, though, *P. macropalpis* uses its inside third leg, hooking into a chunk of the ground with its tarsus, and then accelerating out of the turn just like the Dark Knight, except smaller, creepier, and less bedeviled by abandonment issues.[33]

Of course, the world's most famous vigilante uses the Batmobile to pursue criminals, his particular form of prey. And that begs the question: What is *P. macropalpis* pursuing?

Nobody knows. But it's almost certainly smaller, and could be even faster, than the mighty little mite.

So when it comes to relative speed, we might not have yet found the world's fastest runner. But there are other members of the "fastest" club that are far more certain.

WHY ENGINEERS ARE TAKING A SECOND LOOK AT FALCONS

People long suspected the peregrine falcon was the fastest bird in the world—and the fastest animal of all, in terms of absolute speed. Its top velocity was long theoretical, though, because falcons operate in a vast, unpredictable, and very three-dimensional environment that makes a good radar-gun fix quite difficult. As recently as the late 1990s, we didn't really know for sure how fast *Falco peregrinus* could fly.

That didn't sit well with Ken Franklin. The professional pilot, master falconer, and amateur scientist knew birds had played an essential role in human flight. The Wright brothers extensively studied avian aeronautics before taking off at Kitty Hawk, and Orville Wright later wrote that "learning the secret of flight from a bird was a good deal like learning the secret of magic from a magician. After you know what to look for, you see things that you did not notice." And yet, Franklin lamented, we didn't even know what the fastest of all birds was capable of, because we weren't looking.[34]

So he decided to find out. And, since radar guns weren't going to get the job done, he decided to take a different road. The high road, as it were.

Others had tried to calculate falcon speed by what was observable from the ground. But Franklin knew that peregrines often soar several miles in altitude—far past heights at which ground observations were possible. So, starting from a few thousand feet and moving progressively higher, Franklin and his falcon, whose name was Frightful, began a training regimen that culminated in both man and bird diving from a Cessna 172 at 17,000 feet.

Franklin wore a video camera. Frightful wore a half-ounce recording altimeter. Pursuing a lead-weighted lure that Franklin dropped once they were both diving together, Frightful tucked into a dive and reached a speed of 242 miles per hour.

At that speed, he was falling the length of a soccer field every second.

Franklin was hopeful the data he and his team gleaned from the Frightful experiments might help aerospace engineers better understand how to reduce drag and turbulence. And he worked hard to convince them to take a deeper look at the peregrine's body shape, wing contour, and feather configuration during high-speed dives.

It turned out to be a tough sell. The writer Tom Harpole, who spent years following the exploits of Franklin and Frightful, thought for sure he'd found someone interested in understanding what airplane makers could learn from birds when he met Jim Crowder. Crowder was, after all, a senior technical fellow at Boeing whose specialty was studying airflow to improve the performance of planes. Crowder was also an amateur birder.

But while Crowder said he believed that "birds do all kinds of things that are unknown and potentially worth finding out about," he also warned Harpole that the aviation industry saw itself as "a mature business" that had moved past birds as sources of knowledge for flight. Crowder lamented the conventional wisdom that, if there were aeronautical discoveries yet to be had, "someone would have found them by now."

"Looking back, I do understand where they were coming from," Franklin later told me. "I didn't have a PhD. These people had spent their whole lives trying to quantify the mathematics of flight and, from their perspective, I was the new guy on the block who was throwing birds out of airplanes."

Frightful passed away around 2012, and Franklin has retired from the skydiving game. He keeps pigeons these days instead of birds of prey. "Frightful and I made more than 200 jumps together," he said. "We took it as far as we could."

For more than a decade after Frightful set the animal air-speed record, falcon freefall got little more than a passing look from the aeronautics set. That finally changed, though, in the early 2010s.

That's when a team of German scientists realized that maybe it wouldn't be such a bad idea to at least take a look not only at how peregrines manage to go so fast, but also at how they withstand the high mechanical loads that push and pull against the birds' 2-pound frames when they maneuver at such speeds. After all, when pulling out of an extreme dive while clutching a lure weighing nearly as much as she did, Frightful was confronted with more Gs than the limit for the US Air Force's F-22 Raptor.[35]

Building from observations taken during Frightful's falls, the Germans trained a group of falcons to dive in front of a 200-foot dam. At that height the birds couldn't reach maximum acceleration, but they did tuck into the same body and wing configuration Frightful had when falling much faster. Because the dam offered a high-contrast background, the researchers were able to reconstruct the bird's exact flight path and body shape using multiple high-speed video cameras. With those images, the team built a life-sized model of one of its falcons, slathered it with oil paint, and put it in a wind tunnel. The streaks of paint showed how air moves around a falcon's body during a fall.

And that's when the German team noticed something interesting: regions of the model along the back and wings where paint had accumulated, indicating a separation of wind flow. When they went back to look at images of their birds, and honed in on that area, they noticed a series of small feathers that were popped up from the falcon's body at the exact same locations the paint had pooled on the models. They hypothesized that the arrangement of feathers prevented the flow separation seen on the model.[36] Somehow, it seemed, the birds knew which areas of their wings were not moving air as efficiently, and had figured out a solution to the problem.

That finding excited Marco Rosti, then a doctoral student at the University of London. The young Italian aeronautical engineer was part of a team looking for novel ways to address the issue of stall, which happens when the direction of an aircraft wing and the direction of the oncoming airflow get too far out of parallel, causing significant airflow separation and loss of lift. The problem is as old as aviation; Otto Lilienthal, a pioneer in glider flight, died in 1896 after a crash caused by stall.[37] The century that followed has given us a tremendous number of innovations in aviation, but we haven't "solved" stall.

Falcons seem to have solved it, though. So, building off what had been learned in the falcon experiments, Rosti and his fellow researchers devised a flap that could be hinged on the top side of a wing with a torsion spring. The self-activated flap was designed to pop up, just like the little feathers on a falcon's wings, to disrupt the airflow separation.[38]

Rosti said that the entire time he'd been studying aeronautics he was told the same sorts of things about animal flight that Ken Franklin had been in the wake of the Frightful flights. "What we heard was that perhaps some animals like insects were good to help us identify completely new ways of flying," Rosti said, "but not for helping us improve the way we already fly."

And yet interest in his team's falcon-inspired solution to stall was red hot—and the enthusiasm was coming not just from airplane designers, but from the helicopter community as well, which also faces that age-old problem, albeit in different ways.

Rosti remains cautious. There are a lot of remaining hurdles, not the least of which is an aviation culture that is wedded to ideas about how airfoils are supposed to work, even when those ideas begin with the premise that, in a lot of situations, airfoils won't work.

Ultimately, Rosti said, he accepts that his team's design might not revolutionize air travel. But if it makes it a little less bumpy for some folks, he said, it will be worth the effort.

Perhaps more importantly, though, the bio-inspired design has proven people like Franklin right. We may be more than a century into our era of aviation, but falcons are millions of years into theirs. The idea that there's nothing more to be learned from birds when it comes to human flight is pure hubris.

It's just a story we've told ourselves. And stories aren't always true.

HOW A WIDELY TOLD FISH TALE HELPED BLUEFIN TUNA CLAIM A SPOT IN THE RECORD BOOKS

As fish tales go, this one's pretty good: From 1908 to 1935, sport fishermen would gather throughout the year at the Long Key Fishing Camp, an ocean anglers' paradise off the tip of Florida made famous by the legendary American writer Zane Grey. And it was there, the story goes, that members of the club once observed a sailfish take out 300 feet of line in three seconds. If true, it meant that particular fish would have been swimming at 68 miles per hour. And that would make *Istiophorus platypterus* the fastest fish in the world.

It's on the basis of this claim that the sailfish is widely credited as being "the world's fastest fish." The 68-mile-per-hour record has made its way onto thousands of websites, has been repeated by reputable publications like *National Geographic* and *Field & Stream*, as well as the National Oceanic and Atmospheric Administration,[39] and appears in quite a bit of scientific literature as well.[40]

The primary source for the alleged observation, however, appears to have been lost to history—and it might not have come from Long Key at all.[41]

The funny thing is, once it was finally put to the test with an accelerometer-based series of measurements, it turned out that 68 might be quite a bit slower than the sailfish can actually go. Researchers at the University of Miami have concluded that *I. platypterus* is actually capable of hitting 78 miles per hour in a very short sprint.[42]

But in improving the sailfish's top speed, the Miami scientists might also have shortened its reign, because when the folks at the University of Massachusetts' Large Pelagics Research Center—better known as "Tuna Lab"—learned about the new sailfish record, they got to thinking. If the purported sailfish record was so much slower than its actual potential speed, they wondered, what else was wrong?

The lab's researchers had been watching Atlantic bluefin tuna, *Thunnus thynnus*, for years, and knew that particular fish, which can grow as heavy as 1,500 pounds, was exceptionally fast. Perhaps, they thought, it could reach instances of acceleration that pushed it past the 45-miles-per-hour mark typically referenced by sources that seemed little more credible than those that previously had been used for sailfish.

Working from the same playbook as their Miami colleagues, oceanographer Molly Lutcavage and her colleagues at Tuna Lab tagged a bunch of bluefins with miniature satellite tags. The tags are often used to track the movements of migrating animals, and are designed to pop off after a month, float to the surface, and send out a signal so that they can be recovered—usually by cooperating fishermen.

When a tag prematurely popped off an 800-pound bluefin after just a week, and the researchers got it back to Tuna Lab, they were stunned. The fish had been swimming so fast that the device had been torn into pieces.

How fast does a fish have to swim to shred a popup tag? The data download suggested the bluefin had reached a top speed of 144 miles per hour.

The tuna had been moving at more than twice the speed we'd long believed was the maximum for a fish.

Just like cheetahs, bluefin don't keep their top speeds for long. The researchers estimated their record-setting tuna had maintained those super speeds for a matter of only a few seconds. But, not for nothing, those speeds were being reached in an environment that is roughly 800 times denser than air.

Remember how disinterested the aeronautical engineering community was in falcon flight? It's early, yet, to know whether the nautical community will react with similar dispassion to the newly recognized masters of underwater speed, but it bears noting that there wasn't much interest in understanding sailfish speed when it was widely assumed those creatures were the fastest things underwater. It wasn't until 2013 that any meaningful examination of *I. platypterus*'s hydrodynamics was undertaken.[43]

But bluefin might be a different story. For one thing, we've already got a better scientific starting point for their superlative claim to fame. For another, they're tuna—and everybody knows tuna. Canned, sesame crusted and seared, or in a spicy sushi roll, we eat a lot of the stuff; it's thus one of the most valuable fisheries in the world.[44]

They're also in trouble; Atlantic bluefin are listed as endangered by the International Union for the Conservation of Nature. The Center for Biological Diversity has petitioned the US federal government to also list *T. thynnus* under the Endangered Species Act, although so far to no avail.

At the intersection of economic value and population decline, there is often a flurry of research attention. That was true for bluefin even before it was suggested that they might be the fastest thing in the sea. Recent years, for instance, have brought studies on bluefin habitats in the Mediterranean Sea,[45] how bluefin DNA is different in different parts of the ocean,[46] and how tuna migrate and aggregate across the oceans.[47]

And then there's the US Navy's GhostSwimmer, a robot designed to look, and swim, like a tuna. The autonomous underwater vehicle was designed as part of a secretive research and development operation called Silent Nemo. It was inspired in no small part by the RoboTuna project at the Massachusetts Institute of Technology, where engineers abandoned more than 200 years of conventional thinking about submarines to build an underwater vehicle designed to operate like a fish. The result was a submersible that was more maneuverable and used less energy than conventional unmanned submarines—and which, the Navy quickly recognized, could blend into the marine environment even better.

While the military hasn't officially said so yet, it seems likely that GhostSwimmer could be a replacement for a frequently criticized Cold War–era operation, the furtive Navy Marine Mammal Program. The military organization uses dolphins and sea lions to hunt for mines, keep watch for underwater infiltrators, and recover lost equipment,[48] and also provides animals for translational research aimed at addressing human diseases and health concerns.[49] But that latter role doesn't sit well with some animal welfare activists, who have pushed for the Navy's program to be closed, reasoning that captive animal research should directly impact the conservation of the animals being studied.[50]

One hang-up to swapping animals for robots: At least for the moment, GhostSwimmer isn't as fast as dolphins and sea lions, nor does it have the endurance or dive capacity that the Navy's trained marine mammals do. The robot fish can move up to 17 miles per hour in short durations or travel for longer periods at 3 miles per hour. A bottlenose dolphin can exceed 22 miles per hour and cruise for long periods at up to 7 miles per hour.[51]

But neither can come anywhere close to the burst speeds of a bluefin. And that could be important in a tactical setting. With that in mind, will the Navy seek to make its mechanical tuna even more tuna-like through a study of the natural adaptations that make bluefin so fast? Almost assuredly.

But will it succeed in making a mechanical animal that comes anywhere close to the 144-mile-an-hour burst speed capacity of the one Mother Nature designed? Probably not.

There's absolutely no doubt that we can greatly improve our technology by looking to how animals have evolved in nature to solve challenges similar to the ones we face. But the more we do, the more clear it becomes:

Nature is way ahead of us.

Chapter V

AURAL SECTS

How Superlative Sound Helps Drive Life as We Know It

I'll never forget that scream.

I'd been on a crowded plane that morning and an even more crowded bus that afternoon. Then it was the rusty bed of an old pickup truck. And then it was the bristly back of what seemed to be an even older donkey. As the sun set over the lowland forests of Ecuador's Esmeraldas Province that night, I rolled out my one-person tent and slid into my sleeping bag.

And then it happened.

The guttural, gravelly roar of the Ecuadorian mantled howler is truly something to behold. It rises above all of the other noises of the forest and carries for miles. But I wasn't miles away. I was right under it.

Despite that experience, I never really wondered what makes howlers so loud. I suppose I just figured that *something* had to be the loudest thing in the jungle, and howlers were it.

But Leslie Knapp sees the world in a different way. And thank goodness she does, because her search for answers as to what makes howlers so aurally extraordinary resulted in one of the most charmingly amusing research findings in the history of biology.

Knapp is a biological anthropologist whose main area of study is the major histocompatibility complex, the same part of the immune system that permits a cheetah to take a skin graft from virtually any other cheetah. Her focus, though, isn't on cats but primates. Knapp studies how histocompatibility genes vary in monkeys, apes, and other simians, and how the diversity of those genes is generated and maintained.

Like me, Knapp had learned about the howler's awe-inspiringly ear-splitting roar during her travels through Central and South America. But she noticed something I didn't.

All of the howlers were loud. Really loud. The loudest land mammals in the world, by some estimates. But some species seemed even louder than others—and that didn't seem to correlate to species size. Knapp wanted to find out why.

Her investigation quickly zeroed in on the howler's hyoid bone, which works a bit like a megaphone to amplify its screech. The golden mantled howler, *Alouatta palliata*, has a hyoid that is just under one-half a cubic inch in volume; its call is among the quietest of its noisy brethren. The South American black howler, *A. caraya*, has a slightly bigger hyoid and thus a deeper, more booming call. The brown howler and Yucatan black howler, *A. guariba* and *A. pigra*, respectively, have even larger hyoids and louder calls. And, at nearly four cubic inches in volume, the hyoid of the Venezuelan red howler, *A. seniculus*, is the biggest of the bunch. It gives the red howler an exceptionally low, loud, and resonant roar, which it uses to show off, call out to potential mates, and try to scare away competitors.

There appears to be a rather substantial trade-off for a large hyoid, though: The red has really small testicles.

Now, because I know you're probably asking: Yes, there is an established method to measure animal testes. It starts by using digital calipers to determine the length and width of each testicle. After that, it's a simple matter of using the common formula to calculate the volume of a prolate sphere.[1] Then, to account for variability that exists between the left and right testes, both numbers are added together to get "total testicular volume."

When Knapp plotted call volume against ball volume, she found the loudest monkey had the smallest testicles, at less than a quarter of a cubic inch in volume. The quietest had the biggest, at a hefty 1.4 cubic inches in volume. The rest fell right onto an inversely proportional plotline.

The loudest monkeys, it appeared, were compensating for something.

Monkeys with smaller testes produce less sperm. For that reason, Knapp figures, they have to work harder to pass along their genes—by getting the attention of more mates.

This was more than a clever, carnal finding. Knapp's howler study was the first time scientists had seen an evolutionary trade-off between sexual physiology and vocal characteristics, and it has given other researchers another framework with which to understand the ways in which the making and processing of sounds—like the calls of the tiny bumblebee bats in Southeast Asia—can drive speciation. That's informed additional studies on the evolution of mice, deer, dragon lizards, and frogs.

Knapp figures it might teach us something about us, too. Howlers and humans are evolutionarily close cousins, after all. When the study was released in 2015, Knapp said the research "helps us understand primate behavior, and that way, we can learn more about ourselves."[2]

Take, for instance, our love for fast—and loud—cars.

There hasn't been a lot of scientific research on "compensation cars." But the year before the "calls and balls" study was published, a British car leasing company reported that it had asked more than 500 luxury and sports car owners—and their partners—some rather personal questions. The results were not flattering.[3]

To be fair, the survey's sponsors don't appear to have had a control group of men who drive quieter, less ostentatious vehicles. And more formal scientific inquiry on the subject tends to show that men are often quite generous in their self-assessments, likely regardless of what sort of car they drive. Even anonymous studies that rely on self-reported measurements, for instance, offer averages that are substantially bigger than those in which researchers get a little more hands-on.[4] Alas, what we project is often different than what we offer. What the calls-and-balls study tells us, though, is that our inclination toward compensation might not just be the result of human culture. There might, in fact, be something deeper at work.

And that's just the start of a cacophony of research related to how creatures that exist at—and well beyond—the limits of human hearing use the aural spectrum not only to communicate, but also to navigate, pursue prey, and evade capture.

HOW THE SIMPLE ACT OF LISTENING LED TO A ZOOLOGICAL REVOLUTION

As is the case with a lot of other superlatives, there's no clear consensus on what creatures are the loudest in the world, because sound is created, transmitted, and perceived in many different ways. What we recognize as "loudness" is a combination of frequency, intensity, and duration. Howler monkeys often get record-holding billing because their cries fit nicely within humans' range of perception for all three of those parameters. They roar at frequencies we can hear, with an intensity that makes our eardrums rumble, and for a long time. A very long time, as I came to learn the first night I camped in the Ecuadorian forest.

It wasn't until quite recently that we paid much attention to animal-generated frequencies outside the human range of hearing, which begins at roughly 20 hertz and extends through about 20,000 hertz.[5] That anthropocentric bias kept us from some rather basic revelations about our animal brethren, and for far longer than it should have.

Even those who had spent most of their lives with elephants, for instance, didn't seem to realize the very narrow range of sounds we could hear weren't the only ones the animals could make. And when that discovery did come, it came from someone who had only just begun listening to elephants.

For decades, biologist Katherine Payne had been working to record and analyze the sounds of humpback whales, resulting in the landmark discovery that whales "sing" to one another using complex refrains, evolving melodies, and even repeated patterns that resemble human rhymes.[6] In the mid-1980s, a colleague from what was then known as the Washington Park Zoo asked Payne to come to Oregon to compare notes on megafauna. She could tell the zoo's staff what she knew about humpbacks; they'd reciprocate with information about elephants. Intrigued, Payne was on a flight to Portland within a week.

Payne wasn't content to just talk about elephants. She wanted to hear them, too. So she asked the zookeepers to give her some time to be in the presence of their largest charges. The keepers obliged, though not before offering a somber warning: Don't get too close. "You'll be noodles if they drag you through the bars," one of the men warned.[7]

As soon as the keepers left, though, Payne found herself surrounded by thick, gray, wrinkled trunks, sniffing her clothes, feeling her shoulders, and meeting her eye to eye. To Payne, it seemed as though the elephants were trying to make it clear that she was welcome in their presence.

Payne left Oregon a week later, knowing she'd had an experience most folks could only dream about. But she didn't think she'd seen or heard anything of scientific importance.

It wasn't until she was midair, on the plane flight home, that it hit her. She'd had a nostalgic feeling, being in the presence of the elephants; it had been hauntingly reminiscent of her teenage years, when she sang in a Catholic choir while standing next to the pipes of the chapel organ. It was something deep and resonant, and by the time she had landed, she knew she needed to get back to Portland.

When she did, she came equipped with an infrasonic tape recorder. And when she sped up the recordings to make them audible to the human ear, the whole world changed.

These enormous animals, creatures we'd known for millennia—or thought we'd known, at least—were so much more complex than we had ever imagined. All along, we'd assumed the only sounds they were emitting were the ones we could hear. As a result, we figured they were mostly quiet beasts—trumpeting when alarmed, grunting a bit here and there when in the midst of something strenuous, but not particularly prone to chattiness. We now know the truth: Elephants are some of the world's great talkers.

Payne's further research in Africa confirmed elephants were communicating with one another all the time, at up to 90 decibels (that's like standing next to an espresso maker when it's churning out a chai latte)—but at frequencies humans can't hear.[8] And owing to the fact that very low frequencies can travel a lot farther than higher ones, they were communicating across great distances, chatting up fellow travelers across long stretches of savannah.

Once we knew how talkative elephants actually were, the floodgates burst open. We realized bull elephants announce they are in musth.[9] We began to see the different roles that short- and long-distance communication play in the lives of these animals.[10] We began to reassess the ways in which elephants had evolved.[11]

The torrent of research that stemmed from Payne's remarkable realization didn't stop with elephants. It inspired other scientists to look for animal vocalizations that were outside the range of human hearing. That led to discoveries about whales,[12] cows,[13] and rodents,[14] which in turn led to research that has informed the ways in which we protect dolphins,[15] care for farm animals,[16] and model human depression, respectively.[17]

That latter work is particularly important for those with mental health challenges. The overwhelming majority of clinical medical research, including a lot of research aimed at better understanding the way our minds are impacted by our lives, is conducted with animal models. But for a long time, we weren't even listening to all of the different frequencies lab animals use to communicate. Payne's work prompted additional research into animal sounds that are outside human hearing, and that gave psychological researchers additional ways to understand the animals in their labs. That has led to a significantly improved capacity to understand mental health in lab rodents, which is vital for our understanding of human mental health.

The frustrating thing about all of this is that we really could have gotten to it sooner. At the point Payne began to suspect there was more to elephant communication than what we could hear, it had been more than 100 years since Francis Galton had proven, through the invention of something we now call a dog whistle, that animals can hear ultrasonic sounds.[18] And it had been nearly a half century since a Harvard undergraduate named Donald Griffin brought a cage full of bats into the office of a professor who had developed a machine capable of detecting ultrasonic frequencies, revealing bats could not only hear but also speak in such tones.[19] If very small animals can communicate above the level of human hearing, after all, it should have stood to reason large ones might be able to communicate below our range of perception. But once we've pigeonholed something, even something as big and loud as an elephant, it's hard to break out of our assumptions.

Today we know elephants aren't alone in their use of very low frequencies to communicate. In Africa, they share the infrasonic airwaves with hippos, rhinos, and giraffes. Out on the savannah, the sounds we can't hear may actually be louder than those we can.

We also now know bats aren't the only small mammals who use ultrasonic communication. Rodents of many species are chirping away at frequencies too high for our ears. Rats, for instance, produce a 22,000-hertz alarm call in dangerous situations, and also have a 50,000-hertz call they use in friendly encounters with other creatures[20]—so high it's even outside the range of hearing of most dogs.[21]

Those revelations have bolstered another assumption we long held about animal calls at the highest and lowest ends of the spectrum, but hadn't tested: that an animal's calling frequency generally correlated to its body size.

That's what Kobe Martin thought. And the graduate researcher at Australia's University of New South Wales told me she would have been perfectly happy to simply continue to accept the bigger-means-lower assumption as a scientific fact, if only there had been some bit of science to cite when she mentioned it in her research. But when she went to find a reference for that widely held principle, she came up empty.

"No one had sat down and looked at it in great detail," she said. "There was this widely accepted concept that large mammals have low-pitched voices and small mammals have high-pitched voices. People had thought about it. But no one had thought to quantify it."

That lack of research wasn't just impacting our ability to understand the calls of other mammals. One of the things that distinguishes us as a species is the way we communicate. Understanding how our abilities to communicate evolved requires an understanding of how other mammals evolved to communicate. A lack of serious study, Martin said, left us unable to understand our place in the hertz-emitting hierarchy, and unable to so much as approach the question of *why* we speak and listen in the frequencies we can make and hear.

"When something is as big and widely assumed as this idea," Martin said, "someone usually comes around and says, 'Maybe we should test this.'"

And since no one had, she did. Martin and her team scoured the scientific literature for the minimum and maximum frequencies of every mammalian species they could find—nearly 200 of them in all. Then they plotted the frequencies of the animals' calls against their body masses, just like the folks from EcoNetLab had done to arrive at their conclusions about how size impacts speed.

At first, everything was looking as expected. There were some outliers—"cheaters," Martin calls them, like the howler monkey, which had evolved specialized equipment for making sounds deeper than their body size would typically allow.[22] By and large, though, the trend was clear: The bigger the animal, the lower its calls.

Then Martin's team got to the aquatic mammals. And that's when the data went haywire.

The researchers knew, of course, that marine mammals like dolphins often have high-pitched calls. Martin said she expected, though, to see a parallel trend in which those calls got progressively lower as the animals got bigger, in line with what was happening with terrestrial mammals. And that would make a lot of sense in the vast ocean, she said, since bigger animals need more space, and low frequencies travel longer distances in the water.

That's not what was happening, though. Some of the biggest animals, like the mighty baleen whales, were among the highest pitched. Some of the smallest, like the adorable eared seals, were among the lowest pitched.[23]

It's important to remember that aquatic mammals all evolved from animals that once lived on land. At one time in their evolutionary history, it can be assumed, all of these creatures were subject to the as-it-gets-bigger-its-voice-gets-deeper principle. But "it's as if the ocean environment released these animals from that rule," Martin said.

So, when it comes to sound, size matters. But environment matters even more. Size may be, as John Bonner suggested in *Why Size Matters*, the greatest driving force for all of biology, but it's not always the force that drives every change.

Martin's work demonstrated specific traits are often the result of an evolutionary game of roshambo.[24] Perhaps more importantly, though, her findings showed us, once again, that superlative outliers can do a great deal of damage to widely held assumptions about how the living world works. And if there's one thing that really sparks great science, it's the destruction of widely held assumptions.

WHY YOU CAN'T CALL EVERY NOISY CREATURE A LOUDMOUTH

We've known about the *Micronecta scholtzi*'s special talent for decades, but it has sort of been hidden in the research.

Back in 1989, a University of Helsinki zoologist named Antti Janssen published a report about the insect's novel noisemaking. "The sound production of Micronecta," Janssen wrote, "involves the rubbing of a ridged area of the basal processus of the right paramere (pars stridens) against one or two ridges (plectrum) located near the median edge inside the pocket formed by the left lobe of the eighth segment."[25]

OK then. Still awake?

I don't begrudge scientists for writing like scientists. But science doesn't have to be stodgy. It can be fun. It can be irreverent. It can be gross. And when you tell people that the rapid, noisy series of chirps they would hear if they were standing next to just about any pond in Europe are, in fact, the result of an insect that is rubbing its penis[26] against its ribbed belly . . . well, they tend to pay attention.

M. scholtzi's booming wang has been recorded at about 99 decibels—as loud as a helicopter flying just 100 feet overhead.[27] Relative to its .07-inch length, this species of lesser water boatman is the loudest-known animal in the world.[28]

This is more than just a funny superlative fact. It's an avenue for opening eyes (and ears) to the different ways sound is created and utilized in our world. Because when we think about what creatures are the loudest, we typically think about which ones create the most decibels with their lungs, throats, tongues, and mouths. But across the living world, animals have evolved very different methods of creating sound-producing pressure waves, and they have very different purposes for doing so as well.

Take, for instance, the tiger pistol shrimp. When it snaps at something with its claw, a plunger-like tooth on the claw's moveable finger, also known as the dactyl, forces water into a socket on the immobile propus, causing a rapid jet of water to shoot out that can incapacitate prey or intimidate other shrimp.[29]

While it was long assumed the loud sound emanating from the action was the result of the claw snapping together, in 2000 a team from the University of Twente in the Netherlands demonstrated through high-speed video recording that the pop actually occurs when the vapor cavity at the end of the water jet collapses.[30] The result is an up-to-210-decibels sound that has been giving sonar-using underwater explorers and warfighters trouble for more than 100 years.[31] The calculations the Twente team made in its analysis of the shrimp's snap, though, is now being used to help scientists who use air guns and hydrophones to map the ocean floor.[32]

It's not just underwater creatures that have figured out how to use their various appendages to cause a racket, of course. Most folks know crickets make their sounds by drawing the "scraper" on one wing, along the "file" on another. The intervals between the resulting sounds have been shown to be a remarkably good predictor of outdoor air temperature, a phenomenon called Dolbear's Law.[33]

You might think the loudest of these crickets would be hard to lose track of, but that's what happened to the Colombian bush cricket, *Arachnoscelis arachnoides*. The spider-like katydid was first described in 1891. After that, entomologists figured the insect had either gone extinct or been mistakenly identified as a separate species in the first place. It wasn't rediscovered until 2012. With a song exceeding 110 decibels, how did it managed to hide so long? Part of the reason is that most of its calls are ultrasonic—we simply couldn't hear them.[34]

Just as herpetologist Christopher Austin proved in the forests of

New Guinea when he discovered the world's smallest frog, sounds can lead us to incredible new discoveries. What else might we discover if we begin to pay better attention to the world that exists outside our range of hearing?

It's time to listen up. After all, our very existence—and the existence of every living thing on this planet—started with a sound: a tiny pop that changed everything.

Today, oxygen comprises 21 percent of the volume of the air in our atmosphere, and it's been like that for a long time. But it wasn't always like that. There was, at first, no free oxygen.

The first "free O" in our world arrived by way of cyanobacteria, which released it as a waste product when it split water to get hydrogen. That process, the first instance of photosynthesis in our known universe, began about two and a half billion years ago, give or take a few hundred million years.

At first it happened a little. And then it started to happen a lot. Light from the sun would be absorbed by a chlorophyll molecule and, in a nanosecond, the molecule would lose an electron and become positively charged. "The result," Paul Falkowski writes in *Life's Engines*, "is that for a billionth of a second there is a positively charged molecule and a negatively charged molecule inside a protein scaffold, and they are separated by only a billionth of a meter."[35]

That's a situation that can't last, because positive charges attract negative charges. And when that happens, the protein scaffold collapses, creating a pressure wave—a tiny popping sound that may have been the first noise ever made by any life-form.[36] Pop by pop by pop, tiny cyanobacteria pumped dioxygen into the atmosphere in what has come to be known as the Great Oxygenation Event, creating the conditions that have permitted life as we know it to exist on this planet.

Long before any creature had evolved to have anything resembling a mouth or throat or lungs, life was already a noisy affair.

And it just got noisier from there. Just not always in the ways we imagine.

WHAT DO CROCODILE SOUNDS TELL US ABOUT DINO DADDIES AND MOMMIES?

I really do hate to damage people's reverence for the mighty *Tyrannosaurus rex*. The indubitable badassness of that dinosaur, after all, has led a whole lot of young people to investigate the world of paleontology, a veritable gateway drug for other biological sciences.

But when I ask students to tell me what the loudest creature ever to walk the face of the planet was, *T. rex* is almost always one of the first guesses. And, well, there's a tiny problem with that.

OK, actually, there's a really big problem with that.

T. rex, you see, was an archosaur, like today's avians and crocodilians. Birds chirp, sing, squawk, and honk. Crocs gurgle and grumble. But neither of those groups of animals roar the way many of us have come to believe these big, hungry, prehistoric carnivores did.

Loud and resonant roars, like those of lions and tigers, don't appear to have evolved in any species outside of large mammals—which didn't come along until long after the dinos were gone.[37] Such sound-making is the product of vocal cords that are flat and square-shaped, which stabilizes the cords and allows them to better respond to air being forced out of the lungs.[38] Dinosaurs, though, don't appear to have had vocal cords at all. What's more, the biological tools that modern archosaurs use to make the sounds they do make—a syrinx for birds and a larynx for crocs—evolved well after dinos went extinct. So forget roaring;

dinosaurs might not have been able to vocalize at all. Indeed, *T. rex* might have been the strong silent type.

If dinos did make noise, research suggests, the sounds they made were likely "closed-mouth vocalizations," low-pitched gurgles similar to those made by larger birds like ostriches and cassowaries, and by all of the crocodilians, including alligators and gharials.[39]

So dinosaurs may not have been big, scary roarers. However, research from an international team of scientists studying archosaur vocalizations suggests they might have been pretty good parents.

First, the researchers recorded the calls of three kinds of crocodile, plus the American alligator and the spectacled caiman of Central and South America, noting how the calls shifted in frequency and pitch as the animals aged and grew. When the calls of small juveniles—those up to about 14 inches long—were played back to these crocodilians, the mama reptiles moved toward the source of the sound. Meanwhile, the calls from large juveniles—those up to about three feet in length—"hardly elicited an approach." When the baby crocs' calls were electronically manipulated to play back at an even higher pitch, the mamas exhibited an even greater tendency to move toward the calls.[40] That excited the research team, because mother birds have been known to respond in similar ways to baby bird calls.

When you see something that birds do, you know what birds do. When you see something that crocs do, you know what crocs do. When you see something that both birds and crocs do, though, you might just be seeing what dinosaurs did, because what we see in both our reptilian and feathered friends, the researchers wrote, is more likely to have been "rooted deeply in the archosaurian evolutionary tree," the branches of which began to separate about 220 million years ago.

An ever-improving fossil record is helping us understand dinosaur physiology. But deriving dino *behavior* from the fossil record is a much greater challenge, to say the least. The more similarities we can find in birds and crocs, the better idea we'll have about not just what dinosaurs looked like, but what they were like in other ways, too—like how they might have sounded. And once we've got a good guess as to what dinosaurs sounded like in general, there's a chance we can make some quality suppositions as to what they sounded like in specific situations—like when they were feeling relaxed or excited.

Will we ever know for certain what dinosaurs sounded like, how they used their vocalizations, and for what purposes? Perhaps not. But the closer we get to understanding more about them—including and perhaps especially how they communicated—the more we'll understand how our own behaviors fit into a much grander evolutionary picture.

WHY THE WORLD'S LOUDEST FROG CHANGED ITS ACCENT

Long before we arrived on this planet, and long before the dinosaurs ruled the Earth, the masters of our world were amphibians. Big ones. Scary ones. Just plain weird ones.

A 200-pound salamander called eogyrinus. A giant-headed river prowler named megalocephalus. A two-foot-long "snake amphibian" called ophiderpeton.

This was during the Carboniferous period, a 60-million-year stretch that began about 360 million years ago and that brought forth upon this planet a multitude of swamp forests, new plant life, and a whole lot of noise.

We'll never be sure what dinosaurs sounded like, because there aren't any dinosaurs left. But we can make some very educated guesses about the Carboniferous soundscape and its choir of ancient amphibians, because our world is still crawling (and swimming) with frogs, toads, newts, and salamanders.

And they are loud. Incessantly and often ear-splittingly so.

Take, for instance, the coqui frog, *Eleutherodactylus coqui*, which is often called the loudest amphibian in the world. The call for which it was named—"koh-kee . . . koh-kee"—has been measured at greater than 90 decibels. And the little frog, which seldom grows bigger than two inches in length, sings and sings and sings some more.

So loud and so unremitting is the coqui that, shortly after it landed in Hawai'i from its native Puerto Rico, likely in a shipment of nursery plants in the late 1980s, it was declared a dangerous invader. The Hawai'i Invasive Species Council decried its "annoying call from dusk to dawn."[41] It was the first and only time I know of that a government institution went to war with a species primarily because of the noise it creates.[42]

Like so many so-called invaders in so many other places across the globe, though, the coqui has proven adept at thwarting those who believed they could stop its conquest. Hawai'i has spent millions of dollars on dozens of schemes to kill off the frog—including weekly volunteer hunting parties set up neighborhood-watch style and one short-lived effort to caffeinate the little amphibian to death.[43] Yet the coqui abides. In some areas, according to the state of Hawai'i, there are more than 10,000 of them per acre.[44]

When she began studying coquis in Hawai'i a few years ago, ecologist Karen Beard was interested in investigating a common concern—that the frogs would compete for food with native birds, reducing populations of endemic species. She suspected they would. But after scouring the Big Island, where the coquis are rampant, meticulously counting both amphibians and avians, and comparing the numbers to historic data, Beard and her team learned the native population hadn't suffered at all. The biggest difference was that the nonnative bird population had *increased*.

"It looks like the coquis were a good food source for them," Beard told me.

Sydney Ross Singer, whose family maintains a 60-acre refuge for the cacophonous frogs on the Big Island, has argued the "Frog War" is nothing more than a witch hunt, pointing out that more than a quarter of the creatures that make up Hawai'i's current population of arthropods aren't native either—and just as nonnative birds eat a lot of coquis, the coquis eat a lot of other nonnative species.

"I'm disturbed by what we're teaching our children," Singer told me. "For a while the school system had a coqui bounty campaign in which children could kill frogs and bring them to school for prizes. We're teaching them it's OK to kill something just because you don't like the sound of it."[45]

The stop-it-at-all-costs approach also ignores the benefits potential organisms like the coqui frog might offer, particularly as it pertains to seeing—and hearing—evolution in action.

Owing in no small part to its decibel level, there are few animals in the world whose calls have been so intricately studied as the coqui. For a half century, starting in Puerto Rico, researchers have been trying to understand not just how the frog manages to be so loud, but *why* it needs to be, and attempting to ascertain how the coqui uses its piercing call within its ecological niche. One interesting finding is that the two parts of the frog's call seem to have distinct purposes. By recording both parts of the call and watching the responses of various frogs as the calls were played as recorded ("koh-kee"), separated ("koh" and "kee"), and reversed ("kee-koh"), neurophysiologist Peter Narins has demonstrated the first note serves to establish territory, while the second note is used to attract mates. Essentially, "koh" translates into "stay off my lawn" and "kee" means "well, *hello* there."[46]

That, at least, is what the calls mean in Puerto Rico. But as the coqui has added more stamps to its passport, we've been getting an opportunity to understand how animal calls can be affected when a creature goes global.

Anyone who has ever heard a friend's accent and vocabulary shift when they move to a new place knows how fast humans can adopt the tones and lexicon of a different part of the world.[47] Lexiconic adoption makes some sense for humans; it signals familiarity with, understanding of, and even acquiescence to the established culture on the part of newcomers. But what would happen to your accent if you moved somewhere where there weren't any fellow humans?

That's essentially what happened when the coqui landed in Hawai'i. Since there weren't native frogs of any species there already, and no animals as loud as it is, *E. coqui* entered an "empty" auditory niche. Theoretically, the lack of competition should have meant the coqui's call didn't need to change. But it did. And fast.

Already, researchers have seen a significant, and quintessentially Hawai'ian, impact. In the Aloha State, the frog's "well, *hello* there" call is as loud and proud as ever, while its "stay off my lawn" call is now significantly quieter. Researchers believe this correlates to population density: In some areas the coqui population is as much as three times denser in Hawai'i than in Puerto Rico.[48]

What this tells us is that animal calls can be quite sensitive to environmental change—an observation backed by recordings Narins took of coquis he found along an eight-mile stretch of Puerto Rico Highway 191 in the Caribbean National Forest, first in 1986 and then again nearly a quarter-century later. During the intervening years, Narins learned, the frogs' calls increased in pitch and shortened in duration—a change that correlated to a significant increase in temperature.[49]

While most of our attention was focused on eradicating it from places it "doesn't belong," the world's loudest frog was calling out to warn us of the impact of climate change.

In our defense, it's not always easy to hear such warnings. We're not moths, after all.

HOW BATS AND MOTHS ARE FIGHTING AN EVOLUTIONARY WAR

Want to take a guess as to what was not in Hayward Spangler's obituary? (If you said "whatever superlative thing he discovered," you're right.)

But it's not just the folks who wrote the University of Arizona entomologist's obit who missed his contribution to our knowledge of *Galleria mellonella*, the greater wax moth—it was also fellow scientists who study it.

In the early 1980s, Spangler collected moth larvae from an infested honeybee comb in Tucson and raised the moth babies in a small room in his lab at the Carl Hayden Bee Research Center. Once they reached adulthood, he blasted the moths with a quick burst of ultrasonic noise he made with a transducer he purchased at RadioShack. Using a laser vibrometer, which allows researchers to measure vibrations without touching a vibrating surface, he found that a pair of tympanic hearing organs on the abdomen of *G. mellonella* were sensitive to sounds at up to 320 kilohertz, or 320,000 vibrations per second.[50] When the moths were flying, Spangler later noted, they would respond to the ultrasonic noises by folding back their wings like a peregrine falcon and diving to the floor, or by looping and landing on the nearest surface—mirroring behaviors observed in moths when bats approach in the wild.[51]

Spangler published his findings in the *Annals of the Entomological Society of America* and the *Journal of the Kansas Entomological Society*. Both were upstanding peer-reviewed publications, but neither was exactly required reading in the wider world of biological sciences. Even today, with the advent of academic search engines with seemingly endless libraries, Spangler's early work on moths can be hard to track down. Compounding this problem, Spangler recorded his observations but didn't put them into context—he didn't ever say, "I have discovered a moth than can hear frequencies higher than any other known animal."

In 2013, a report from researchers from the University of Strathclyde in the United Kingdom did just that. The study, which was published in the much-more-well-known journal *Biology Letters*, was praised by one fellow scientist for offering a "shocking increase in the frequency sensitivity of moths' ears" that would "require researchers to rethink" the rules of auditory systems.[52]

What did this "shocking" study show? Pretty much what Spangler had already shown: that a pair of tympanic hearing organs on the abdomen of *G. mellonella* were sensitive to sounds at 300 kilohertz.

There was one difference: The new study included the words "the highest frequency sensitivity of any animal."[53]

Framing is everything. Among the media organizations that jumped on the superlative story—without noting Spangler had made a similar discovery thirty years earlier—were the *New York Times*, the BBC, and *National Geographic*.

Spangler is hardly the first scientist to have his or her discoveries overlooked by other scientists and the wider world. Today just about everyone knows Gregor Mendel as the father of modern genetics. His work demonstrating heredity, though, was roundly ignored during his lifetime and not "rediscovered" until three other scientists, Hugo de Vries, Karl Correns, and Erich von Tschermark, reached similar conclusions. That didn't happen until nearly a half century after Mendel's now-famous experiments with smooth and wrinkled peas, and some sixteen years after his death.

Setting aside the notion of who first recognized greater wax moths could hear at supremely ultrasonic frequencies, however, we can focus on a fascinating mystery: How did *G. mellonella* even come to own this aural niche in the first place?

Not much in nature happens at 300 kilohertz. Not that we know of, anyway. *G. mellonella*'s chief predators are bats, whose highest known echolocation calls taper out around 200 kilohertz. So why does the greater wax moth need to be able to be able to pick up more than 100,000 extra vibrations per second?

One possibility: Wax moths are to bats as pronghorns are to cheetahs. That is to say, while the wax moth doesn't need to be able to hear at 300 kilohertz *these days*, it might not always have been that way. Fossil records indicate long-extinct bat genera like *Icaronycteris* and *Palaeochiropteryx* had inner ear cavities that were very large relative to their skulls, as do all modern echolocating bats.[54] If any—and certainly if many—of the bats that have crossed evolutionary paths with *G. mellonella* and its predecessors had higher-frequency calls, it would have put evolutionary pressure on the moths to "hear higher."

Another possibility is that the greater wax moth's hearing is the latest procurement in a bat-versus-moth evolutionary arms race of epic aural proportions. This mega-anna-old fight has produced tiger moths, which make ultrasonic noises to "jam" the signals of their predators,[55] and barbastelle bats, which counterbalance their high-frequency echolocation calls with low-amplitude "whispers" moths can't hear until it is too late.[56] By evolving hearing that goes well past its current need, moths not only have the ability to react to lower predatory pitches faster, but also have built in some wiggle room in preparation for the bats' next evolutionary adaptation—a case, perhaps, of preemptive adaptation.

However it happened, this evolutionary battle could impact not just moths and bats, but *us* as well. At the University of Strathclyde, where that second study establishing the upper limit of wax moth hearing was conducted, an electrical engineer named James Windmill is taking what he has learned about insect ears and applying it to the creation of ultra-small, simple, and task-flexible acoustic systems. Among his bio-inspired designs are microphones variably sensitive only at selected frequencies, which could give humans better capacity to filter out noises they don't want to hear and hone in on the ones they do, just as moths must do given their extreme range of frequency perception. The obvious application for such technology is hearing aids, but the technology could also be applied to tiny medical devices capable of detecting very specific signals of stress inside the human body. Similarly, such microphones could be used in industrial systems, offering safety engineers another way to monitor noisy, complex workspaces for signs of trouble that, if addressed quickly, could improve worker safety and stop production delays.

It may have taken three decades after Spangler's discovery of its extreme range of hearing for *G. mellonella* to finally be seen as useful to other scientists. But given the wax moth's history as a pest—it gets its name from the fact that its caterpillars love to munch on beehives and can devastate entire colonies of bees—we might consider ourselves lucky we didn't figure out a way to eliminate it entirely before we could better understand the lessons it could teach us.

That, after all, is what we very nearly did to another aural wonder of the world.

WHAT SPERM WHALES ARE TELLING US ABOUT LISTENING TO THE WORLD

When the first scientific treatise was written about *Physeter macrocephalus,* the general consensus was that it was a rather quiet beast.

"The sperm whale is one of the most noiseless of marine animals," wrote Thomas Beale, who served as the ship's doctor on two whaling ships, the *Kent* and the *Sarah & Elizabeth*, in the early 1800s. "It is well known among the most experienced of whalers that they never produce any nasal or vocal sounds . . . except for a trifling hissing at the time of the expiration of the spout."[57]

Beale, of course, never observed whales underwater, where they spend the vast majority of their lives. He only saw them when they were either about to be killed or were actually in the process of being killed. Such is the problem with trying to do science as part of a hunting expedition—and this, we would do well to remember, is how much of the natural world came to be described in the eighteenth, nineteenth, and twentieth centuries, as whalers, anglers, trappers, and other killers of animals reported back to scientific bodies their observations about the creatures they encountered, and often destroyed, during their adventures. The starting point for our understanding of thousands upon thousands of species were observations taken while the animals were literally fighting or fleeing for their lives.

But Beale could not have been more wrong in his assertion about the sperm whale's quiet demeanor. Thus his lament that those who had earlier attempted to describe *P. macrocephalus* "should rather have left a blank in the page" seems rather ironic.

Beale did make some rather important observations. But he also got a lot wrong, including his estimation that the sperm whale was probably "the largest inhabitant of the globe." It is indeed a mighty beast, but only about half as long as the blue whale.

The idea that the sperm whale was a "quiet giant" stuck longer than the idea that it was the largest. It wasn't until the late 1950s that research unconnected to the whaling industry was conducted,[58] and revealed that sperm whales do, in fact, make noise. Those noises are mostly clicks and buzzes, as opposed to the "songs" many of us associate with whales, but *P. macrocephalus* was anything but the "noiseless" creature Beale had described.

It would be nearly a half century further down the road before scientists got around to measuring those sounds. When they did, they found the sperm whale's rapid-fire "clicks" regularly exceeded 200 decibels, and a team of researchers from Denmark's University of Aarhus recorded one whale at 236. In absolute terms, that was "by far the loudest of sounds recorded from any biological source."[59]

Why does a sperm whale need to be this loud? Consider the environment in which it hunts, up to 6,000 feet deep in the blackness of the ocean. Its tiny eyes don't do it a heck of a lot of good down there.

That's where its huge head comes in. The whale's noggin, which takes up a third of its body, is full of hundreds of gallons of a waxy substance known as spermaceti. (For some reason, people once thought this stuff was whale semen, hence the creature's name, but it's really just a water-insoluble chemical compound of fatty acids and alcohol molecules.) Twisting through the spermaceti are two nasal passages, one of which runs to the whale's blowhole and the other of which runs to an organ known as the phonic lips, which some scientists call "monkey lips" because that's what they resemble. When the whale smacks these lips together, the sound reverberates through the spermaceti, a number of air sacs within it, and the animal's skull. This all happens in a matter of milliseconds.[60]

And that's when the game of Marco Polo begins.

For the tragically uninitiated, Marco Polo is a summertime rite of passage, a swimming pool game of tag in which the player who is "it" closes their eyes while other swimmers flee. When the blind swimmer says "Marco," the others must say "Polo," giving the pursuing player a clue as to where they have gone.

That's how sperm whales catch squid. Except instead of screaming "Marco," the sperm whale pops those monkey lips together and then waits for the echo to bounce off those tasty cephalopods. And, of course, for the whale—and especially the squid—it's not a game.

Wanting to get a better look at this, a team of researchers led by biologist Patrick Miller from the University of St. Andrews in the United Kingdom headed to the northern Mediterranean Sea and the Gulf of Mexico, where they tagged twenty-three whales with devices to record the animals' sound, depth, and orientation. Almost all of the whales' rapid-fire clicks, also known as "creaks" or "buzzes," came in the deepest part of their dives—and correlated to times of intense maneuvering.[61]

That's absolutely remarkable, because squid are quite squishy—not exactly the sort of objects one imagines would "bounce sound" all that well. What sperm whales must lock onto are the squids' parrot-like beaks, which at their largest are just a few inches long, but are usually much smaller.

The US Navy has long studied bats and dolphins, searching for clues about how those animals use echolocation, nature's tremendously more advanced version of human-made sonar. That research has led to advances in what the military calls "environmentally adaptive target recognition"—the ability to easily and flexibly filter out background noises by using multiple frequencies to "see around" distracting sounds, like those pesky snapping pistol shrimp.[62] That's especially important for locating small objects in the water—a life-or-death imperative in a world in which underwater mines are becoming smaller and easier to manufacture.

The sperm whale offers another model for echolocation—one based not only on frequency and amplitude, but also on the rhythm and patterns of its clicks. So far, though, there has been very little research dedicated to better understanding the way sperm whales echolocate, and how we might be able to emulate them. The world's largest and loudest echolocator—and potentially the best as well—has gone virtually unstudied.

For now, the Navy's chief interest in sperm whales comes as a result of the controversy surrounding how warfighter-based sonar disturbs, and even kills, marine mammals. It's no secret that the US military's environmental record isn't good, and defense leaders have fought hard against environmentalist lawsuits and court orders seeking to protect animals sensitive to the tremendously loud noises the Navy pumps into the oceanic acoustic environment. If naval leaders realized how much they stand to gain by protecting *P. macrocephalus*, they might change course. And if the inner workings of a unique echolocation system built by millions of years of evolution isn't enough, perhaps naval researchers will be drawn to another newly emerging area of research related to sperm whales: The animals also might have a very finely tuned magnetic navigation system.

The way researchers came upon this hypothesis is quite sad. As far back as the Middle Ages, people have been documenting whale beachings, but although we've long known whales sometimes strand themselves, and sometimes do so in groups, scientists aren't sure why. One of the most promising, albeit tragic, opportunities to study this phenomenon presented itself in early 2016, when twenty-nine sperm whales, all of them male, were found on the beaches of Germany, the Netherlands, Great Britain, and France. Autopsies were conducted on twenty-two of the whales, and it appeared that they were all in good health before getting stranded.

Given where they were found, the fact that all of the whales were male was not particularly surprising. Females and young tend to stay at lower latitudes—they typically get no farther north in the Atlantic than the Azores Islands, west of Portugal. But when males reach independence, around the age of ten to fifteen, they form bachelor groups that migrate much farther north.

But young male sperm whales do this every year. So what was different in 2016? Solar storms—coronal mass ejections that play havoc with Earth's magnetosphere, subtly disrupting the magnetic polarity of our planet. There were two storms around the time of the beachings, and researchers from Germany and Norway, who published their findings in the *International Journal of Astrobiology*, believe the storms could have caused the whales to become disoriented.[63]

If that's true, it means sperm whales don't just navigate by echolocation, but also with the help of the Earth's magnetic field, possibly through trace magnetic elements buried inside the whale's spermaceti. Essentially, this theory goes, the whales navigate for long distances by "listening" to the magnetism of the Earth—perhaps in ways that can inform our own navigational technologies.

We very nearly didn't learn any of this, though, because sperm whales, like so many of their fellow cetaceans, were almost hunted to extinction. By the time Herman Melville published *Moby-Dick* in 1851, the widespread whale slaughter was well underway—and was already besieging global whale populations. Melville romanticized, excused, and diminished the scope of the slaughter, writing that the sperm whale was "immortal in his species, however perishable in his individuality." By the time Melville died in 1891, the estimated global population of about 1.1 million sperm whales had been cut by a third—and that was before the advent of industrialized whaling. By the time the global ban on whaling took effect in 1986, that population was down by two-thirds.[64]

If the ban had not taken effect, there is little doubt we would now live in a world whose loudest creature had fallen completely silent—long before we realized how much we had to gain from the simple act of listening.

Chapter VI

THE TOUGH GET GOING

How the World's Strongest Organisms Might Lift Us to the Heavens

Had Michael Cooney told me he was working with a live velociraptor, I still wouldn't have been so excited.

Don't get me wrong: Velociraptors are completely cool. Especially the real ones—not those big, gray, scaly ones that chase and eat people in the movies, but the "fluffy feathered poodle from hell" paleontologists Stephen Brusatte and Junchang Lü say actually existed in the Cretaceous period.[1] Now those were some tough critters.

But the operative word is "were." Dinosaurs may have been hard-core in their day, but when fate delivered a meteoric wallop to the Earth, dinos didn't make it. They weren't tough enough.

Oh, hey, sure. Lots of things got wiped out then. But lots of things *didn't*, too. And some of those things are still with us today.

Take, for instance, the creatures Cooney was working with the first day I walked into the Sinclair Lab at Harvard Medical School. "So basically," Cooney explained, as he hunched over an array tray with a pipette in his hand, "I'm taking tardigrades and—"

"Wait," I interrupted. "You've got water bears?"

"Well, yeah . . ."

"Can I take one home as a pet?"

"Well . . . I . . ."

I was just kidding about the pet thing. Mostly at least. But my enthusiasm for tardigrades was completely genuine.

It's hard, though not impossible, to see a water bear with the naked eye. The largest only grow to about a millimeter and a half in length. But under a microscope, they're charismatic and downright cute.

I know. Creatures with translucent skin, eight stubby legs, clawed feet, puffy faces, and circular mouths that look like the Sarlacc from *Return of the Jedi* probably shouldn't be described as "cute." But with their chubby skin folds and squinty eyes, somehow they seem downright cuddly.

More to the point: In addition to being the perfect combination of ugly and adorable, tardigrades are the toughest animals on the planet.

I'm usually a bit more circumspect about superlative declarations. There are always objective caveats, after all. And there are lots of ways to measure toughness—in terms of absolute or relative force generation, the ability to kill or avoid being killed by other animals, or the capacity to pull, push, or carry a large load a long way. But if we stick to the most basic definition of toughness—being strong enough to withstand adverse conditions—it's awfully hard to argue that tardigrades aren't unequivocally the world's biggest badasses.

Water bears have evolved very little over the past 500 million years, braving cataclysmic meteor showers, ice ages, severe shifts in the gaseous makeup of our planet's atmosphere, and a ton of other things that have wiped out entire orders of animals, taking whatever Mother Nature (and human beings) can throw at them like the embodiments of so many Chuck Norris jokes.[2]

But even though tardigrades have been known to scientists since before the founding of the United States, and are found throughout the world, including some of the highest mountain peaks and deepest ocean trenches, they weren't the subject of much serious study until researchers started doing all these horrible things to them—and realized, as they did so, that there's likely a lot we can learn from these tough little critters.

Because in a world in which our future is anything but assured, there's a lot to be gained from studying the world's best survivors.

On the one hand, if there's a phylum that gives lie to the myth that humans are going to completely destroy all signs of life on Earth, it is the tardigrades. Stephen Hawking was probably right when he said our days on this planet are numbered,[3] but plenty of animals won't even shrug when we depart. Fellow theoretical physicists David Sloan and Rafael Alves Batista, who ply their trade at Oxford, just down the road from Hawking's old office at Cambridge, have built mathematical models showing that even billions of years of events far more calamitous than the temporary advent of *H. sapiens*—cataclysms like asteroid strikes—are unlikely to push the mighty water bear into extinction.[4]

"Life, once it gets going, is hard to wipe out," Sloan said. "Huge numbers of species, or even entire genera, may become extinct, but life as a whole will go on."[5]

So what could kill off the tardigrades? Perhaps nothing, Sloan and Batista have suggested, but our own sun bloating into a red giant and consuming the Earth.[6]

In the same way that Alan Weisman's *The World Without Us* paints a hopeful picture of the future—even, and perhaps especially, if humans don't survive much longer—Sloan and Batista's work offers a window into a world, hundreds of millions if not billions of years in the future, in which our planet isn't some silent, barren rock spinning around a star, but a place where life abundantly goes on.

That's the good news.

The bad news is that, even if a whole lot of water bears will indeed make it through the human-caused Holocene Extinction, not every tardigrade species is likely to survive. The Antarctic tardigrade, *Acutuncus antarcticus*, may be exceptionally well adapted to live in one of the harshest climates on our planet, but it may not be able to adapt fast enough to survive the increasing temperatures and ultraviolet radiation that are befalling the bottom of the planet as a result of global warming.

When scientists from Italy's University of Modena subjected *A. antarcticus* to heightened levels of radiation and rising temperatures, in line with some predictions about what will happen in Antarctica in coming years, they found that even members of a phylum of animals widely extolled for their toughness might be in for a losing fight. A lot didn't make it, and the water bears that did survive produced fewer eggs and were delayed in reaching sexual maturity.[7]

So no, we probably can't wipe out all of the world's tardigrades. But yes, our actions can impact some of the strongest, most resilient, and most evolutionarily steadfast creatures on our planet. And that should make us sit up and take notice. For if our actions can adversely affect the toughest creatures of our world, just imagine what we're doing to ourselves.

Tardigrades might also be able to show us ways that we can make ourselves tougher. And, in doing so, they might help us ensure our survival—both on and off this planet—at least for a little longer.

HOW TARDIGRADES CAN HELP US REACH OTHER PLANETS

There was a time, biologist Paulo Fontoura told me, when tardigradologists were considered "dreamers" who studied "useless animals."

Fontoura, who researches tardigrades at Portugal's University of Porto, has discovered many species of tardigrade. But until quite recently, he said, such discoveries were rather inglorious work done for no greater purpose than developing a better understanding of the vast spectrum of life on this planet.

These days, Fontoura said, tardigrades' emerging reputation as survivors has helped many scientists realize water bears "may contribute to our understanding of biological phenomena like aging, cancer, and more, with future potential uses in medicine." But, he noted, none of those applications would have been possible if there weren't scientists willing to study these organisms before they were known to be so damn tough.

Even now, the process of describing new species of tardigrade "does not bring big grants, many citations, or publications in high impact journals," Aslak Jørgensen, an editor for the journal *Zootaxa*, told me. But those discoveries, he said, are "the fundamental building blocks for understanding life on Earth." Every new publication helps increase that understanding, he said.

"Tardigrades are not flagship species for biodiversity—although they are very cute—but they are omnipresent and their extremotolerance makes them very suitable for public outreach," Jørgensen said.

That's especially true in the case of *Ramazzottius varieornatus*, the most widely studied of more than 1,000 tardigrade species and, at least so far, the toughest of the bunch.

Put them in boiling water? No problem. Freeze them at temperatures near absolute zero (almost –460 degrees Fahrenheit)? They're good. Blast them into outer space? Rock on. Shower them with radiation? No biggie. Dehydrate them like an emergency food kit? Just add water. Freeze them for decades? They can survive that, too, through a process known as anhydrobiosis, in which they dehydrate to the point of carrying only 1 to 3 percent of their water weight, shrinking into a semi-living form called a "tun" while they wait for conditions to improve.[8]

Takekazu Kunieda has made it a good part of his life's work to find out why tardigrades are so punk rock, and in 2015 he and his team at the University of Tokyo took a major step toward answering that question. When they sequenced the genome of *R. varieornatus*, they found tardigrades have evolved a variety of genetic strategies for staying alive when times get tough. Among the most fascinating of the team's findings was that this creature lacks several very common gene pathways promoting stress damage, like the ones activated in humans when we step into the sun.

Less stress doesn't mean no stress, though, so these tardigrades have extra genes that help fix DNA damage, and have evolved to produce a novel protein that works like a "radiation umbrella" for their DNA.

After that discovery, Kunieda and his teammates did what any of us would do if we knew how: They incorporated the tardigrade protein into human cells and exposed them to radiation. Those cells suffered 40 percent less damage under X-rays.[9] That could have significant implications for interplanetary space travel, where radiation exposure is a significant concern for astronauts on years-long trips to other worlds.

Michael Cooney was building on that research the day I met him at Harvard, where he was a postdoc studying cell-based therapies to delay or reverse aging. He was introducing genes from tardigrades into human cells to see if those genes would improve repair for other sorts of DNA damage—and he was excited by what he was already learning about two genes that seemed to be able to protect human cells exposed to hazardous chemical compounds.

Given its remarkable formidability, it makes sense *R. varieornatus* has gotten so much recent attention from researchers. But there are a lot of other water bears out there, and new ones are being discovered all the time. In 2017 alone, scientists discovered a new tardigrade species in the soil sediment on the border of streets in Chetumal, Mexico; three new species on the Brazilian coast; five species in the mountains of Colombia; two on the Mediterranean island of Sicily; and two more living on the lichens growing on the rocky Western Atlantic shores of Portugal and Spain.

It's possible, I suppose, that *R. varieornatus* will turn out to be the toughest of the bunch. It's far more likely, though, that one of these new species—or a tardigrade yet to be discovered—will prove to be an even better survivor.

What might happen if we introduce genes from those tardigrades into human cells? The potential for a transgenic "toughening up" of our genome is virtually endless.

And that's just one phylum of very tough little creatures. There are plenty of other superlative survivors whose genomes we're only just beginning to study.

HOW THE WORLD'S SLOWEST-EVOLVING ORGANISM IS HELPING PRESERVE BIODIVERSITY

Some folks call it a ghost shark. Others call an elephant shark. Whatever you call it, the Australian ratfish known as *Callorhinchus milii* is a marvel of our underwater world. For while the oceans have changed quite a bit in the past 450 million years, *C. milii* hasn't changed much at all.

To put this into context: In the immediate aftermath of our divergence from *C. milii*, our human ancestors were animals of the sea that had just recently developed backbones and were on the verge of developing limbs. We hadn't even yet evolved into the famed Tiktaalik, the four-legged "missing link" paleontologist Neil Shubin helped discover and wrote about in his beautiful book, *Your Inner Fish*.[10] In the years since, the branch of the tree of life we're on has split and split and split to give us mice and cows, possums and chickens, lizards and frogs, and even the fish that—before the sequencing of *C. milii*'s genome—was believed to be the slowest evolving vertebrate in the world, the human-sized, bottom-dwelling coelacanth.[11]

But the ancestors of the elephant shark at that time? They likely looked a heck of a lot like the modern version of the elephant shark.

When underwater videographer Pang Quong got a call from some researchers in Melbourne about a few elephant sharks that had been caught and studied, and that were now due for release, he knew he had a chance to witness something truly special in the wild. Elephant sharks are typically deep sea fish, but for a few months each year they come closer to the shores of Australia and New Zealand to spawn. "But when turning up near Melbourne, elephant sharks seem to prefer mud flats and areas with extremely poor visibility," Quong told me. "So when I got the opportunity to see a lot of them released back to the wild I just had to get in the water with them."

Quong waited by a pier as the researchers set the fish free in a shallow, clear shelf, then followed them as they headed into deeper waters, making one of the only video records in the world of the way elephant sharks swim. "They use their pectoral fins like birds flying," Quong recalled. "In my daily routine I get the opportunity to dive with many animals and fish, from whales to sea dragons, but this dive was one of those special moments."[12]

Until fairly recently, though, it might not have felt very special at all. That's because to most people, elephant sharks were just another fish in the sea—notable only for the fact that they were commonly diced, fried, and served with chips in New Zealand.

Then Byrappa Venkatesh came along.

The geneticist from Singapore's Institute of Molecular and Cell Biology didn't choose to sequence the elephant shark because he suspected it was old, but instead for a very practical reason: The fish's genome was relatively short.

Thus far, there has been little rhyme or reason for how scientists choose which species they will sequence. Frequently researched species like fruit flies, roundworms, and lab mice have all been through the process. Domesticated animals like cows and sheep have, too. Hundreds of thousands of individual human genomes have been sequenced.[13] But outside of that, it's been quite haphazard, with promising leads, individual hunches, and personal or political interests guiding decisions.[14]

As a chairperson for the international Genome 10K project, Venkatesh is part of a group of scientists that includes Stephen O'Brien, who discovered the genetic uniformity of cheetahs, and Emma Teeling, who was part of the research group that established the divergence of call frequency in bumblebee bats living in Thailand and Myanmar. Hoping to bring a little more order to the process of genetic cataloging, the group wants to sequence an animal from each of the vertebrate genera—about 10,000 species in all. That project was launched at the University of California, Santa Cruz in 2009 after the scientists agreed that, even though the cost of sequencing a new animal was still quite high at the time, and although the process still took a long time, they really couldn't wait any longer.

"We are losing biodiversity so quickly," Venkatesh told me. "We knew the cost would go down eventually, but it was important to get started right away."

Some cartilaginous fish have genomes larger than the human genome, while *C. milii* has a genome about one-third the length of *H. sapiens*. That made it a good starting point for the project.

First, though, they had to get a good sample. "The fish we sequenced is one I caught," Venkatesh proudly said of a trip to Tasmania, where he hired a fishing guide who led him to a popular spot for ratfish. "We caught several, actually. It was quite an exciting day." Back onshore, Venkatesh dissected the fish, separating out tissue samples from the brain, gills, heart, intestine, kidney, liver, spleen, and testis, the latter of which was initially used to generate an initial sequence before a more accurate gene annotation was generated through RNA sequencing.

By comparing the protein sequences in all known genomes and comparing those to the known fossil records, researchers have established rates of divergence that allow them to estimate a creature's evolutionary age. When Venkatesh and his colleagues plugged the elephant shark's genome into that algorithm, they were shocked.

"There were some indications that shark genomes might evolve slowly—sharks have very low metabolic rates and there is evidence that this can be used as a proxy for the pace of evolution—but I didn't expect anything like what we found," he said. "I would call it a very pleasant surprise that, for the elephant shark, not much has changed for a very long evolutionary period, and that makes it a very useful reference—now we can use it to compare to other vertebrates, including humans, to better see and understand what changes have occurred, when, and why."

"But even though the elephant shark hasn't changed much in a half-billion years, haven't we changed a lot?" I asked. "Wouldn't there come a point at which we'd diverged so much that there's not a lot we can learn about our own genome?"

Venkatesh laughed. "What we've found is very interesting," he said. "If we look at humans and elephant sharks and dogfish and fugu, which is a pufferfish, what we find is that there are more changes between the fishes than there are between humans and sharks. Looking at conserved non-coding elements, between humans and fugu there are a couple thousand. When we do a similar analysis with humans and elephant sharks, we find more than 4,000."

For instance, the team's work revealed elephant sharks and humans share a gene that has been the subject of considerable evolutionary speculation: the tumor suppressor known as p53, the same gene at the center of inquiries into why elephants are so incredibly cancer resistant. Cartilaginous fish are also cancer resistant, although likely not as much as elephants. But a detailed comparison of the elephant shark's p53 to other vertebrates shows it has evolved more radically than other genes from the same family of protein coders.[15] This means it may do its job in a markedly different way in *C. milli* than it does in *L. africana* or *H. sapiens*—yet another opportunity to better understand our shared genetic potential.

But even the vast areas of the genome where there is no overlap can offer us tremendous insights. For instance, researchers have also discovered the elephant shark completely lacks several genes vital to many other animals' immune systems—including the protein-coding gene known as *CD4*, without which humans would be left vulnerable to a wide range of diseases that would inevitably wipe us all out.

That finding challenges common notions about how biological defense systems fight disease. We humans tend to think of pretty much every element of our physiology, including our immune systems, as being quite advanced. Yet the elephant shark is capable of mounting a sophisticated immune response—one that has kept it merrily swimming along the bottom of the ocean for hundreds of millions of years longer than humans have even walked upright—without the additional genetic tools we've developed for keeping sickness at bay.

Having more genes doesn't mean having better genes. Sometimes, it turns out, less is more. And that understanding could help us uncover, in our species' shared genes and others as well, novel approaches for addressing human diseases.[16] Indeed, from the slowest evolving vertebrate we've yet discovered, we might be learning a few tricks that will help in our own survival.

But, then again, sometimes less isn't more. Sometimes more really is more.

HOW AN ANIMAL WITH THE WORLD'S LONGEST KNOWN GENOME MIGHT UNLEASH OUR X-POWERS

The first time I saw an axolotl, I was in elementary school. I was at a pet store buying a fish, which I was planning to train like a porpoise after watching the dolphin show at a marine-themed amusement park near my home.[17]

A few years later, a teacher asked my biology class about the strangest animal we'd ever seen. I couldn't remember the name of the creature I'd seen at the pet store, so I tried to describe it. "It was pink, like bubblegum," I told the class. "And it had a smiley face. And a mane like a lion. And gills like a fish. And legs like a lizard. And a tail like a tadpole."

I got made fun of a lot that week. And for years to come I doubted whether I'd actually seen the animal or merely dreamt it up.

But the axolotl is real. And, despite its squishy, happy, and altogether adorable appearance, it's really tough.

Science writer Kristin Hugo[18] once compared the axolotl, also known as the Mexican walking fish, to a character from a superhero comic book.[19] And, indeed, it's got a lot more in common with the Marvel Comics antihero Wolverine than *actual* wolverines do because, just like the most famous of the X-Men, it's a master of regeneration.

Like many lizards and amphibians, it can regrow its tail. But, like a much smaller number of salamanders, it can also regrow its limbs. And its skin. And its jaw. And its eyes. It can even mend a severed spine. And it can do all of this again and again, dozens of times over, with no scarring and no loss of quality in whatever piece of its body has been restored.

Scientists have been intrigued by *Ambystoma mexicanum* since the English zoologist George Shaw first described it in 1789, but they've long struggled to understand its most famous attribute. They took a big step toward solving the mystery, though, in 2018 when they published the axolotl's genome, which at 32 billion base pairs is the longest known sequence in the world. That's ten times as long as the human genome. And hiding in all that code is the secret to regeneration, one of the holy grails of medical research.

It might not be so sought-after if it seemed unlikely that humans could ever regenerate. But here's the thing: We already do. When our skin grows back after being cut—that's regeneration. Our ability to regenerate is limited to our skin and, to a lesser extent, our livers, but that ability comes from the same DNA that produces all of the other tissue in our bodies, from our bones to our muscles to our internal organs to our brains. In theory, all those other tissues could regenerate, too.

Could the axolotl help us unlock the secret to our own X-powers? Maybe, but only if we can keep it around.

Given how easy it is to find a Mexican walking fish in pet stores—and even meat counters—around the world, you might not suspect it is critically endangered in the wild. Because they're quite hearty, easy to care for, and easy to breed, have the world's largest amphibian embryos[20] (which make them ideal for stem cell research), and are, quite frankly, very cute, there might be hundreds of thousands of them in farms, pet shops, home aquariums, and research laboratories around the globe.[21]

But in the series of waterways in southern Mexico City that are the last known wild habitat for *A. mexicanum*, the population of axolotls has plummeted in recent years, from 6,000 per square kilometer in 1998 to just 35 per square kilometer in 2015, due to declining water quality, wetland depletion, and the overpopulation of nonnative fish that eat the walking fish and its eggs.

Randal Voss, who contributed to the 2015 study showing the precipitous population decline and to the 2018 gene sequencing report, is the keeper of a collection of thousands of axolotls at the University of Kentucky. He has acknowledged, though, that his stock is inbred, which "can compromise the health of a captive population."[22] What's more, over time, creatures kept and bred in captivity will be selected for fitness in captivity, and what that will do to genes that have evolved for life in the wild is anyone's guess.

Ecologists have spent decades working to protect the remaining wild walking fish, with little success. The completion of the sequencing study might be the push that's needed to finally get worldwide attention—and funding—flowing to Mexico. That's not simply because the genome is so uniquely long, but because when they searched through all of that code, the researchers couldn't find a gene called *PAX3*, which is active in neural crest cells and was thought to be necessary for the development of bones and muscle tissue.

Up to that point, the assumption was that vertebrates without *PAX3* wouldn't be able to survive. And yet the walking fish walks, thanks apparently to a gene from the same family, *PAX7*, which swooped in to take over some of the functions of its relative.

All of this means the axolotl's greatest contribution to science might not simply be helping us discover the secret to regeneration; it might be showing us a new model for life itself, one in which no gene is truly mandatory to a creature's biological existence and in which the function of any distinct sequence of nucleotides can be bypassed by other sequences. If that's true, it may mean we can build into our genomes an attribute we know is key to toughness and survivability in human-built mechanical systems: redundancy.

Of course, if you were looking at an axolotl for the first time, you might not suspect any of this. It doesn't look all that tough—not in the sort of mean, muscular, thick-skinned way we commonly associate with toughness.

But in that respect, it's certainly not alone.

WHY THE WORLD'S SLOWEST MAMMAL IS ALSO ONE OF THE TOUGHEST

The vine was just hanging there. Right in the middle of the jungle. Right in front of a big, dark pond of water.

It was about an inch and a quarter in diameter and, when I gave it a good, hard tug, it felt as sturdy as any rope I'd ever used to hang a hammock, climb a rock wall, or moor a boat. I grabbed hold with both hands, took a few steps back, and lifted my feet from the ground.

The vine held true, but Tarzan I ain't. I made it only about three quarters of the way over the water when my momentum began to slow. I reached out for another vine, and found the transition trickier than all those movies make it look. The water was only thigh deep, but the mud beneath it was thick, and I had to reach down to pull my boots back onto my feet a few times as I slogged toward dry land.

My guide, a local hunter named Piero Martín, swung past me as gracefully as a spider monkey, laughing all the way across the pond.

Piero and I were trekking through the floodplain forests of northwestern Peru to get a glimpse of the world's slowest mammal.[23] And on the other side of the water, amidst the broad, bright green leaves of a stand of cecropia trees, we found what we were searching for.

"When many people think about the sloth, they think it is very lazy," Piero told me. "But just look!"

Piero pointed up to the crook of some cecropia branches. From my vantage, it looked as though an old soccer ball had been wedged between a limb and the trunk.

I waited. And I waited. And I waited.

"Is something going to happen?" I finally whispered.

Piero began laughing hysterically. "No," he said, slapping me on the back and cackling some more. "Probably not. Sloths are lazy. We could come back tomorrow and there is a good chance that sloth will still be right there, in the same branch, positioned in the same way."

A few days later, while fishing in an Amazon River tributary, I caught my first glimpse of a sloth actually doing something. It had been motionlessly hanging on the thin upper branch of a tree about 20 yards away when I arrived. Now an afternoon storm was moving in, the wind was picking up, and the tree was swaying back and forth at a good clip—enough that the sloth seemed to decide it was time to seek a more stable place to wait out the gale.

Estimates of three-toed sloth speed vary quite a bit, so I applied a bit of middle school math to figure out how fast this one was moving. There were roughly two yards between the branch the sloth was on and the next one down. The sloth cleared that distance in just under sixty seconds, for a speed of .06 miles per hour. And that's moving *with* gravity.

From tardigrades to elephant sharks to Mexican walking fish, animals often seem strange until you can see them in the context of their environment and evolutionary history. But I left Peru more confused by natural selection than I'd ever been before. Because—*somehow*—the three-toed members of the genus *Bradypus* and their two-toed-and-just-a-bit-faster cousins from genus *Choloepus* have managed to survive in Central and South American forests *packed* with carnivores like anacondas, jaguars, and harpy eagles.

Not only that, but these animals have managed to withstand these conditions for tens of millions of years, surviving predators, environmental changes, and other pressures that wiped out a whole lot of other animals, including no small number of other now-extinct species of sloths. Out of the sloth's evolutionary lineage it wasn't the 10-foot-tall mighty *Megalonyx* that survived, nor the hippo-like *Glossotherium*, nor the gorilla-like *Hapalops*, but rather a small group of seemingly hapless tree-dwellers that really don't seem well adapted to fleeing, fighting, or even so much as getting down from a tall tree when a storm comes.

I'm certainly not alone in my confusion about the sloth's suitability for survival. In the mid-1700s, the French naturalist Georges-Louis Leclerc, Comte de Buffon, wrote that the sloth was a "strange and bungled conformation," opining that "one more defect would have made their lives impossible."

Shouldn't natural selection weed out the weak?

Lucy Cooke thinks so. And that's exactly why she believes sloths are anything but weak. Rather, the zoologist and National Geographic Society explorer thinks the sloth is one of the toughest animals around precisely because it's so slow.

"There is nothing defective about the sloth—it is, in fact, a very successful animal," Cooke wrote in 2013. "In tropical jungles sloths make up as much as two-thirds of the mammalian biomass, which is biology speak for 'I'm doing rather well, thank you very much.'"[24]

Wildlife ecologist Jonathan Pauli, who studies sloth biology and behavior at the University of Wisconsin–Madison, agrees. He told me I needed to readjust the way I was looking at the sloth's life cycle. "Rather than thinking about everything that eats sloths," he said, "think about what sloths eat."

They eat leaves—which aren't exactly a superfood in terms of healthy calories. Wanting to understand more about how sloths use the small amounts of energy they get from their primary food source, Pauli and his colleagues collected ten three-toed brown-throated sloths and twelve of their slightly faster cousins, the two-toed Hoffman's sloth. The researchers injected the sloths with a traceable isotope and then released them. About a week later, they re-collected the sloths[25] and checked their blood for the presence of the isotope, from which they could calculate the animals' metabolic rate. When they did, they learned the three-toed sloths were burning just about 100 calories per day, roughly the equivalent of a single tablespoon of peanut butter. That meant they had the lowest metabolic rate of any known mammal. Only the much-larger giant panda comes anywhere close to that low level of energy consumption and output.[26]

While most animals spend their lives looking for food, the tree-dwelling sloths are surrounded by it. "Leaves are ubiquitous but are a super cruddy food," Pauli said. "So sloths have evolved a series of odd physiological, digestive, behavioral, and anatomical characteristics to limit energetic expenditures."

And with their food needs well taken care of, sloths can focus on other things—like making more sloths.[27] Sloths only have one baby at a time and stay with that offspring for years. Over the course of their relatively long lifespans of up to thirty years, however, three-toed sloths can indeed do "rather well" in terms of population success. From the standpoint of species survival, they can actually afford to get picked off, occasionally and even frequently, because there are plenty to go around.

So what lessons do sloths have to tell us about survival? The easy takeaway is that we could all stand to chill out and slow down a bit. The more important lesson, though, has to do with food security: Sloths evolved to eat what is immediately around them and copiously available.

We evolved in that way, too. Far too many of us just haven't realized it yet.

WHY BEETLES, WHICH RARELY GO EXTINCT, COULD HELP FEED THE WORLD

I wasn't in the Siem Reap city market for more than five minutes when it happened. I try to be as inconspicuous as possible when I travel, but I get distracted easily and I'm clumsy, and that combination doesn't lend itself to unobtrusiveness.

I was studying a basket of dried fish when a little boy stepped in front of me and began filling a small plastic bag with tamarind pods. I stepped backward to give the kid space—and felt my flip-flopped foot hit something squishy. An anguished squawk rose above the market clamor, and I spun around to see that I'd stepped directly into a chicken pen.

My family keeps hens, usually three at a time and sometimes as many as eight, so I know my poultry. I bent down to examine the bird, which had one leg wired to the side of the cage. Given the crude way in which it was cuffed, I worried I might have broken its leg, but it didn't seem worse for the wear. Still, its owner wasn't happy with me.

That is how I came to own a chicken in Cambodia.

I didn't keep the bird for long—just the time it took to accept it from the hands of the woman at a cost of 20,000 Cambodian riel, and then to present it back to her as a gift. She was tickled, and offered me a small bag of what I at first took to be roasted nuts. When I looked closer, though, I saw she had given me a bag of mealworms.

"Oh," I said, reaching in to fish out a few of the bugs. "You want me to give the bird a snack?"

I leaned down to feed the chicken, and the woman starting laughing.

"*Ot yl tae!*[28] No . . . to this way," she said, putting her hand in front of the bird.

Then she gestured as though she was eating. "Yes . . . to this way." For emphasis, she grabbed one of the worms from my bag, threw it into her own mouth, and made an exaggerated example of chewing and swallowing.

Crispy and sticky, the bugs tasted like roasted pumpkin seeds cooked in something sweet. When I pointed to the tip of my tongue and smiled, the woman showed me a bottle of Coca-Cola she had poured into the pan. I was eating Cambodian American fusion food.

When I tell this story to many of my American friends, they often look a bit nauseous. When I tell just about anyone else, they look bored. Mealworms are beetle larvae, after all, and beetles are the most widely eaten insect in the world, with about 350 species on the global menu.

This makes sense, because scientists have discovered and named more individual species of beetles than any other animal order.[29] One in four described animals is some form of beetle. The shortest, at one-hundredth of an inch in length, is *Scydosella musawasensis*, which was discovered in Nicaragua in 1999 and not seen again until 2015, when it was rediscovered on a fungus in Colombia.[30] The largest, at more than half a foot, is the Hercules beetle, *Dynastes hercules*, which can also be found in Colombia. There are 380,000 different species of beetles, and we're in no way near done with finding new ones. In 2014 alone, researchers in Indonesia found nearly 100 new species.[31]

So how did beetles become so prolific and diverse? By being evolutionarily tough. Beetles simply don't go gently into the good night of extinction.

That surprising finding was the result of a comprehensive examination of more than 5,500 beetle fossils. When Dena Smith from the University of Colorado and Jonathan Marcot from the University of Illinois compared the fossils to extant beetles, they found about two-thirds of the beetle families that have ever lived—including those that are nearly 300 million years old—are still around today.

Smith, a paleoentomologist, believes the secrets to the success of the myriad members of order Coleoptera are their abilities to metamorphose and move. Soft and squishy larvae can thrive in different habitats than those most comfortable for exoskeleton-clad adult beetles. As such, a short-term but drastic change, like a fire or flood, that wipes out all the beetles in one state of metamorphosis might not impact those in a different state. And since adult beetles are good fliers, they can move quickly in response to longer-term changes in climate, too.[32]

Another factor Smith thinks is pertinent: what beetles eat. Some eat roots, stems, and leaves. Others eat seeds, nectar, and fruit. Some eat living animals. Others consume dead animals. Quite a few chow down on dung. As lovers of the world's forests are coming to understand all too well, some voraciously feed on wood and bark, leading to an epidemic of tree loss made even worse by increasing drought conditions caused by climate change. And many species that might prefer one form of food are more than happy to switch to another when needed.

Beetles just aren't picky about their food. And that brings us back to those mealworms I chowed on in Cambodia. If we're going to stick around as a species for much longer, with a growing global population, we're almost certainly going to need to find another food source—one that requires less land, creates less pollution, and converts plants into protein with far greater efficiency than cows, pigs, and poultry. And to do that, the United Nations' Food and Agriculture Organization believes, we're also going to need to get less picky—and start eating insects more widely.

About a quarter of the world's population already eats insects. And members of the order Coleoptera like the june beetle and rhinoceros beetle—and, yes indeed, even the dung beetle—have some of the highest protein content of any insect. Beetles are among the only insects eaten in the larva, pupa, and adult stages, which means chefs have a lot of different textures and flavors to work with, even within a single species. And given the number and diversity of species to choose from, the beetle buffet is ripe for exploration and experimentation.

That's not to mention all of the other insects. There are some 200 million individual insects out there for every single human on the planet.[33]

When I put the word out on social media that I was interested in finding some local chefs who were working with insects or insect larvae, though, the reaction was a bit disappointing. "I would also like to have this information," one friend responded, "so I can avoid those establishments." Another friend simply responded with a puking emoji. At my local market, the butcher said she was open to the idea of keeping any kind of animal protein that would sell, "but I just can't imagine a world in which insects are anything more than a novelty."

I was ready to give up on the notion that beetles and other bugs could ever be successfully introduced into the mainstream American diet. Then three things happened.

The first was that I reread "Consider the Lobster," in which David Foster Wallace irreverently expounds on the Maine Lobster Festival, where thousands of people gather to boil living animals and eat their remains, paying top dollar to do so. The irony of the expensiveness of lobster is that, through the 1800s, it was "low-class food, eaten only by the poor and institutionalized," because taxonomically speaking, Wallace observed, "lobsters are basically giant sea insects."[34]

The second was that my daughter dressed up like a sushi roll for Halloween in a costume she designed herself and tailored with my mother's help. Until very recently, the idea of eating strips of raw fish was gag-inducing for many Americans; now just about every town with more than a few restaurants has one that serves sushi, many supermarkets have an on-site sushi chef, and kids dress up as sushi for Halloween.[35] My daughter wasn't even the only sushi roll in our neighborhood. She was so proud of her costume that, the week after trick-or-treating, she wanted to wear it again to go out for dinner. The restaurant that night was packed and the staff and fellow diners were delighted by a walking sushi roll that matched the ones on their plates. The foods we're "fans" of can change rather quickly.

The third was that I learned from a friend about Daniella Martin, a proud entomophagist—that is to say, a bug-eater—and the author of *Edible: An Adventure into the World of Eating Insects and the Last Great Hope to Save the Planet*. Martin has also discovered that a lot of people can be downright wimpy when it comes to the idea of eating insects. But the children she introduces to insects-as-food aren't like that at all. "The kids aren't eating bugs for the shock effect," she wrote. "They're not doing it for the photos they can post to their Facebook pages. They're doing it because the bugs taste good and because the idea that bugs are bad hasn't solidified in their as yet unsocialized, unossified minds."[36]

With all that in mind, I stopped by the school where my wife teaches and asked if I could take a quick survey of her third-grade students.

"If I had a bowl of insects and told you they're perfectly safe to eat and yummy, too, who would eat them?" I asked.

The kids didn't blink. Nearly every hand went up. And one girl whose hand stayed down approached me later to say she wasn't worried about eating bugs; she thought I was testing them to make sure they knew not to accept food from strangers.

I returned the next day with a bowl full of baked crickets. Surely they'd balk, I thought. And a few did—but most stayed the course. The ones who said they'd be interested in eating a bug were indeed interested, and gulped them down with little fanfare. A few more who had said they wouldn't try insects the day before ended up giving it a go. Not everyone liked them, but nobody gagged or spit them out. I don't imagine I'd have the same success with Brussels sprouts.

After school, my daughter and I took some insects around for the other teachers to try. Only a few of them were willing to eat the crickets, and generally only after they saw my child do it first. "They taste like crackers," she told one teacher as a bug leg hung from her smiling lips.

The animals that have been most successful, the ones that have evolved to take up the most biomass on our planet, are almost certainly going to have to play a big role in our food security future. And if our kids are any indication, there's little reason why that needs to be seen as a huge hurdle to overcome. Just about everyone can find a bug to enjoy eating—or a bunch of them. We've just got to get a little tougher about trying new things.

WHAT ANTS CAN TEACH US ABOUT BEING SUPER

When I met him, John Humberto Madrid was running around the Amazon jungle with no shoes on.

That's the first thing I noticed. I'm not sure why. Especially since, when I looked up, he had a squirrel monkey hanging around his neck.

One part Doctor Doolittle, one part mad scientist, and two or three parts the grandfather everybody wishes they had, Humberto is the director of Bioparque Del Amazonas, a research station in the jungle of southern Colombia dedicated to caring for animals rescued from traffickers or wounded by hunters, and to providing a place for researchers to come and study the incredible biodiversity that exists in the Amazon, from anacondas to manatees to tarantulas to jaguars.

Part of that biodiversity is a half-inch-long insect called Burchell's army ant; as we darted about the center, Humberto stopped suddenly to scoop up a swarm from the ground. After brushing away most of the insects, he honed in on a single specimen of *Eciton burchellii* he found particularly interesting. Holding it by the abdomen between his thumb and forefinger, he asked me to come closer.

"See those pinchers?" he said. "These animals might have the strongest grip in the world. If a human had equivalent strength he could crush a refrigerator between his arms. Except these are not the ant's arms; they are his mandibles."

By way of example, Humberto put his other index finger right up against the ant's bulbous, peach-colored head and let the insect clamp down on his skin. The sharp ends of the pinchers clamped into his flesh. The tip of his finger turned bright pink.

I winced. Humberto didn't.

"*¡Díos mio!*» I said. "Doesn't that hurt?"

"Oh yes," he said, as though I were asking about the weather. "It hurts very much, as a matter of fact. But not nearly as much as a bullet ant."

"I . . . take it you know from experience?"

He gave me a look that said, *I wish I didn't.*

If there's a Punch Like George Foreman Award in class Insecta, it's got to go to *Paraponera clavata*, whose sting delivers an immediate and overwhelming burst of pain that sends grown men and women to the ground. That sting also gets a score of "four-plus" on the Schmidt Pain Scale, a four-point measurement system developed by Justin Schmidt, who has subjected himself to hundreds of insect stings over his long career as an entomologist at the University of Arizona. Schmidt described the sensation as "like walking over flaming charcoal with a three-inch nail embedded in your heel."

As Humberto pried the army ant's pincers from deep inside the tissue of his finger, a process that looked even more painful than the original pinch, he said he believed ants were among the most tragically understudied organisms in the world.

"This one has a strong grasp," he said. "Others hurt more. But there are many species of ants in the Amazon, and each one is more extreme than the others in some way."

There are more than 12,000 known ant species worldwide. The one thing they have in common, Humberto said, is that they are tougher together.

If you're outside and look around, right now, it probably won't take too long to spot an ant. And once you spot one, it's incredibly unlikely you'll spot *only* one. Ants may be fast, strong, and really well armored, but individually they're relative weaklings. Only together are they strong enough to withstand adverse conditions.

And ants are together a lot. They don't do much of anything alone. They forage together. They fight together. They build together. They navigate long journeys over rugged terrain together using a "scent trail" of pheromones, with each individual ant contributing to the path in a body-odor baton relay. And, because of all of this, they survive and thrive together.

Small wonder ant colonies are often considered to be a superorganism—a word coined by Harvard entomologist William Morton Wheeler to describe a collective in which, rather than evolving to lean on one another to ensure individual survival, the individual subjugates itself to ensure the survival of the whole. The term "superorganism" frequently is used to compare multicellularity, in which cells work together for the good of an organism, and sociality, in which individuals work together for the good of the colony. But while that concept is a convenient metaphor, it's not a scientific principle, and in nearly a century, it had hardly been examined.[37]

There was another century-old idea, though, that biologist Jamie Gillooly and his colleagues figured might be helpful for examining whether colonies really *did* function as singular organisms. In 1905, Austrian physicist and philosopher Ludwig Boltzmann had noted the fight for existence "is a struggle for free energy available for work." The bigger an organism gets, of course, the more energy it needs to expend to do anything. Over the years, researchers have discovered the rate at which energy is expended correlates to an organism's size with remarkable consistency, a principle that has come to be known as the Metabolic Scaling Theory.

If colonies were functioning as organisms, Gillooly believed, their metabolic data would help show it. So he and his collaborators examined colonies of 141 different species of ants, as well as 27 species of bees, wasps, and termites, collecting information on everything from the colonies' weight to their food consumption to the total biomass of their collective gonads. Among other predictions, Gillooly's research team had hypothesized that the colony's biomass production and lifespan would scale predictably to its mass and metabolic rate, in line with the way singular organisms do.

And it did—so much so that, if the charts showing the correlations were not labeled, it would be difficult to guess which data points belonged to individuals and which to colonies. The so-called superorganisms were following many of the same biological rules for energy use as individual organisms.[38]

A few years later, researchers from Bristol University in the United Kingdom showed a colony of *Temnothorax albipennis* ants will make a quick and coordinated response to a simulated attack, but in different ways based on which part of the colony is being attacked. When they attacked a colony by removing a few ants from the nest, the researchers found that the ants evacuated the nest. But when they took away a few ants who were scouting, the colony withdrew back *into* the nest. The ants, the researchers concluded, were reacting just like a singular organism would—moving their "body" when needed, but simply withdrawing an "appendage" when that was all it took to stay safe from harm.[39]

During the devastating 2017 floods in Houston, millions of people around the world got a look at one of the many strategies ant superorganisms use to survive, as photographs depicting giant flotillas of fire ants swarmed social media. Craig Tovey had seen that behavior before. As both a systems engineer and a biologist, the Georgia Institute of Technology professor had long been fascinated with the way ants work together to create structures for survival and protection. By itself, a single member of the species *Solenopsis invicta* won't make it long before succumbing to rising waters. But long before the 2017 floods, Tovey and his colleagues had collected fire ant colonies from around Atlanta, taken them back to his lab, and dropped them into a pool of water. In response, the ants quickly weaved their bodies together in a pancake-shaped mass that the researchers learned could float for weeks.

By freezing the ant rafts with liquid nitrogen, Tovey's team was able to examine the ways in which the ants link together—with a combination of mandibles, claws, and the adhesive pads of their feet.[40] Fire ants do this in minutes, without any sort of training or preparation. Just as most vertebrates, including nearly every mammal, naturally knows how to swim,[41] fire ants naturally know how to keep their colony from drowning.

Doing so, of course, may mean individuals must sacrifice themselves for the good of the colony. When humans commit such acts, we call it bravery and selflessness; we marvel at its audacity and we lament its rarity. But when ants do it, we call it "being ants."

It's hard, when considering what ants do to keep the colony going, not to be inspired that perhaps—just perhaps—humans could learn to be more like them. Indeed, given the way we're increasingly connected to one another, we might be moving in that direction. That's certainly what the brilliant mammalogist and nature writer Tim Flannery has come to believe.

"Is it possible that the invention of the Internet is leading to a similar social evolution of our own species?" Flannery once mused.[42] "As we strive to avert a global economic disaster or agree on a global treaty to prevent catastrophic climate change, we inevitably build structures that, as with the ants, allow the superorganism to function more efficiently." But Flannery also acknowledged the possibility "that our destructive path will catch up with us before we can make the transition."

Humberto, though, likes to think otherwise. "Look around this jungle," he said. "There is so much biodiversity. But in order to evolve to be alive right now, everything here has survived a great deal of chaos. We had to do this, too. I believe we are stronger than we often think. And if we're just willing to learn to live in harmony with all of the life that is around us, we're going to be just fine."

Chapter VII

DEADLY SERIOUS

Why the World's Most Efficient Killers Are Such Effective Lifesavers

The truck was stuck.

Not just a little stuck. I've been a little stuck before and I know what that looks like. The truck was a lot stuck. It was so-deep-in-the-mud-that-I-can't-see-one-of-the-back-wheels stuck. It was we-aren't-going-anywhere-soon stuck.

On either side of the roadside, the savannah grass was nearly a foot over the top of the roof rack, and so thick that I couldn't see more than a foot into it.

And that's when something dawned on me.

It had been just a week since a German architect had disappeared. It had happened just a few miles from where I was standing, blankly staring at the spinning tires and flying mud near the border of Ethiopia and South Sudan.

His truck had gotten stuck, too. And he wandered away for help. And then he was gone. Lions, it was presumed. Or maybe a leopard. There aren't many big cats left in this part of the world, but they're there. And often hungry.[1]

A scout with a gun was awaiting us at a local ranger station. But we hadn't gotten to the ranger station yet. We didn't have a gun. We didn't have anything.

"How ironic would it be," I asked the interpreter with whom I was working, a Will-Smith-as-the-Fresh-Prince doppelgänger who called himself Rico Jinka, "if I got eaten by a lion while writing a book about the world's deadliest animals?"

"Ironic?" Rico asked.

"Funny," I said.

Rico stared into the grass for a moment, swallowed hard, and scrunched up his sweat-covered face.

"That wouldn't be funny," he said. "If you get eaten, we all get eaten."

Our party of four—Rico and me, plus his friends Eramis and Berechet—redeployed in and around the truck. Eramis and I pushed and pulled. Rico threw rocks under the wheels. Berechet put the gas pedal on the floor and worked the wheel back and forth. The mud smelled like death. The truck lurched forward and I fell backward in front of the left fender. The truck missed me by an inch or maybe two, bathing me in foul water and slimy mud as it passed.

I only later realized the real irony. I was worried about lions. But the creatures far more likely to kill me weren't lurking in the grass. They were the tiny organisms lying in wait in the mud. And they were the much larger organisms working alongside me to extricate the truck.

In a good year—that is to say a good year for them, not us—lions kill 100 humans worldwide.

Mosquitoes kill 725,000 people a year, mostly via malaria, which runs rampant in that part of Africa. Other tiny creatures such as tsetse flies, roundworms, and the parasite *Trypanosoma cruzi*, which is carried by the aptly named assassin bug, kill tens of thousands more.

But humans are even deadlier, and that's just to other humans. Nearly 500,000 people are intentionally killed by other people each year. Another 1.33 million die in motor vehicle accidents[2]—like getting run over while trying to get a truck out of the mud.

I was freaked out about the wrong things.

But "deadliest," of course, is subjective in all sorts of ways.

Population size matters. There are about 20,000 lions remaining in Africa—and quadrillions of malaria-causing mosquitoes. When it comes to deaths per individual killer, lions are deadlier.

Size matters, too. Mosquitoes are a lot smaller than lions. When it comes to deaths per pound of killer, the insects rule the day.

How fast does someone or something die after being attacked? How many deadly attacks can one creature accomplish in a day, or over its lifetime? How toxic is its poison or venom?

And then there's the question of "deadly to whom?" Most animals, after all, aren't nearly as deadly to humans as they are to the animals they normally prey upon.

No matter how you measure, though, one thing is clear: The animals we often worry about—"movie scary" creatures like spiders, snakes, and sharks—aren't anywhere close to as dangerous to us as we have been led to fear.

There isn't a household in the United States where you won't find a spider—there's almost certainly one watching you right now—but spider bites account for just seven American deaths per year, on average.[3] You are four times more likely to be killed by lightning.[4]

I've never ceased to leap away when I cross paths with a snake on a hike, and my skin crawls for hours afterward, but in the United States my odds of being killed by any kind of reptile this year are about 1 in 50 million. I'm about 5,000 times more likely to be killed in a car crash, but I don't get the heebie-jeebies every time I get into my Mazda.[5]

Australia spends millions of dollars each year erecting and maintaining shark nets to protect popular beaches, using a "national interest exemption" to get around environmental laws that otherwise would ban the nets, which regularly trap and kill whales, sea turtles, and seals.[6] That's despite the fact that sharks account for just one death per year in a nation of 24 million people.

Far less important than what we should truly be scared of, though, is what we can learn from organisms that have proven, in one way or another, to be prolific killers.

And there are a lot of killer organisms out there.

HOW THE WORLD'S DEADLIEST POISONS ARE LEADING US TO CANCER CURES

The square-shaped acrylic containers were each about the size of a tissue box. There were three of them on the table before me and inside each one, submerged in embalming fluid, was the decapitated head of a lamb.

And if that's not gross enough, these weren't just any lambs. These were cyclopean lambs, with a central eye socket and a fleshy tube projecting from their foreheads. I leaned down to stare into the vacant eye of the one closest to me and released an involuntary shudder.

"That's . . . so . . . cool," I told my host at the US Department of Agriculture's Poisonous Plant Research Laboratory, research plant physiologist Dan Cook.

What I really meant by "cool" was "creepy," but I didn't want to dissuade Cook from showing me any of the other things he kept hidden away in his lab, where scientists work to advise agriculturalists worldwide on the deadly things growing all around them.

Cook didn't disappoint. Soon we were standing inside what might be the world's most dangerous herbarium, a room full of metal cabinets, each one stuffed to the brim with folders holding dried stems, leaves, berries, and flowers of poisonous plants from around the world.

I briefly thought of my favorite character from the Harry Potter chronicles, "herbologist" Neville Longbottom. "Oh my God," I thought, "what Neville could do with a room like this."

What muggles can do with a room like this is pretty amazing, too.

When an animal is poisoned, but a farmer or rancher doesn't know how it happened, they can send a sample from the afflicted creature's stomach to the lab. The lab connects the sample to the chemical signatures of the plants in these files, then sends photos of the offending plant back to the farmer or rancher so they know what to watch out for and remove from their land. "There aren't many labs like this left in the world," Cook said, "so we get asked for help from people all over."

Among the frequently pulled files are those for plants that seem to have come straight out of a Hogwarts spell, like black chokecherry, *Prunus virginiana melanocarpa*; grassy deathcamas, *Zigadenus gramineus*; and greasewood, *Sarcobatus vermiculatus*.

And then there's the folder marked "*Veratrum californicum*," inside of which are samples of the plant's broad, oval leaves and Star of David–shaped flowers. When pregnant sheep eat *V. californicum*, they get a dose of a steroidal alkaloid known as 11-deoxojervine, which scientists also call "cyclopamine" for the effects it has on the lambs of poisoned ewes.

We've known since the 1960s what cyclopamine does to sheep, but decades passed before we knew why: The chemical inhibits a signaling pathway that tells embryonic cells how to go from a little zygote to a big complex organism. The result? Really messed-up sheep that don't usually survive long after birth, if they make it that far at all.

Cyclopamine is one of the most dangerous plants in the world to sheep. But Cook and his colleague, rangeland management expert Jim Pfister, told me it's not a universally dangerous chemical.

"When we're talking about toxicity in plants, there are a lot of variables," Pfister said. "How much of it is available? When is it available? What time of the year is it toxic? There are plants in which toxicity decreases during some times of the year and increases in others. And then, of course, it depends on what animal ingests it. What's poisonous to one isn't poisonous to all."

Plants can be acutely poisonous, meaning humans and animals that come into contact with them get sick or die immediately. This includes plants like water hemlock, *Cicuta douglasii*, which the lab has declared to be "the most violently toxic plant that grows in North America." Just a small amount of the toxin it carries, cicutoxin, can move into the central nervous system in seconds and can cause violent convulsions, grand mal seizures, and death.[7]

Plants can also be chronically poisonous, meaning that they'll only cause harm if ingested over a long period of time. That's what happens with locoweed, a name given to certain species of the *Astragalus* and *Oxytropis* genera. The moniker means what you think: Animals that eat locoweed exhibit signs—like blank staring, extreme nervousness, self-imposed isolation, and violence—that make them seem crazy.

Plants can also be harmless to the animal that eats them, but deadly to that animal's kin. That's the case with the needles of the Ponderosa pine, *Pinus ponderosa*, which at a height of up to 270 feet also happens to be one of the tallest trees in the world. Cattle often graze on the needles when other forage is lacking due to snow, or incidentally eat it when grazing near a tree. And that's generally fine for the cows, even pregnant ones, but bad for the calves pregnant cows carry—the diterpene acids in the needles are an abortifacient.

Similarly, plant poisons can be transmitted from mother to offspring via milk. That's the situation with Madagascar ragwort, *Senecio madagascariensis*, which carries toxic alkaloids that can accumulate in the livers of animals like horses and cause hepatic disease in a foal, even if the mother was unaffected by the poison.[8]

It's the job of lab scientists like Cook and Pfister to work out the permutations for as many poisonous plants as they possibly can. And while their primary mission is agricultural, the "spinoff benefits" of such research are pharmaceutical.[9]

Take, for instance, cyclopamine. The gene that causes embryonic chaos in sheep also plays a vital role in the development of several forms of cancer in humans. Researchers from Johns Hopkins realized in the mid-1990s that if cyclopamine could interrupt that gene in sheep, it might also be able to stop the development of those cancers in humans.[10] Scientists have had a hard time synthesizing the chemical since that time, though, and *V. californicum*, which grows best in high-elevation meadows and streams, has proven a challenge to grow on farms.[11] That's why the drug company PellePharm recently contracted with the US Forest Service to harvest the roots of the plant in the Manti-La Sal National Forest, which, as it happens, is just about 70 miles north of the gigantic aspen forest Pando. PellePharm is using the cyclopamine it harvests to produce a test treatment for Gorlin syndrome, a highly malignant basal cell cancer. What started with mutant sheep may end with hope for tens of thousands of people around the world.

It took more than a half a century from the discovery of *V. californicum*'s toxic properties to reach the point that it was being exploited for pharmaceutical use. These days, though, scientists are increasingly looking to poisonous plants for secrets to combating a wide range of human diseases.

"It's likely that every poisonous plant in the world has secondary compounds, in small amounts, that could be very beneficial to human health," Pfister said.

That's true of true hemlock, *Conium maculatum*, which was used to put prisoners to death in ancient Greece. It's been used in traditional medicine for sufferers of breast cancer, although, like many traditional remedies, that treatment was widely ignored by modern scientists. That finally changed in 2014, when a research team from the University of Kalyani in India showed an ethanolic hemlock extract also had the ability to induce apoptosis through the regulation of p53—the same gene that elephants use to kill off cells that have mutated in malignant ways.[12] The plant used to execute Socrates may wind up saving a lot of other people's lives.

Another killer that could become a lifesaver is deadly nightshade, *Atropa belladonna*, which some scientists believe would have been in the potion that brought death "like an untimely frost / Upon the sweetest flower of all the field" in Shakespeare's *Romeo and Juliet*. It's been used for centuries by assassins, and its alluringly plump and shiny purple berries are occasionally to blame for the deaths of children. But one of its many chemical compounds, atropine, has been used for nearly as long as an anesthetic. Atropine is still in demand across the world today as the standard antidote for nerve gas poisoning, a tragically frequent occurrence even though nerve agents have been illegal under international law for decades.[13]

The deadly castor bean, *Ricinus communis*, is the source of the poison ricin, exposure to which can cause vomiting, diarrhea, seizures, and death.[14] Ricin has been used in a number of modern terrorist attacks and political assassinations, but is also a superstar "phytoremediator" recently proven to be exceptionally effective at extracting from contaminated soil toxic metals like cadmium, lead, and actinium,[15] and chemical contaminants like hexachlorocyclohexane and DDT.[16]

While well-known for their deadly qualities—and increasingly for their pharmaceutical and industrial upsides—hemlock, nightshade, and castor beans are not the most dangerous plants in the world. Not when it comes to human body count, at least. In fact, they're not even close.

But similar to those three poisons, the photosynthetic eukaryote that holds that deadly distinction may also have tremendous potential for helping humanity. If, that is, we can tap into it.

HOW A DEADLY REPUTATION HAMSTRINGS ONE OF THE WORLD'S MOST PHARMACEUTICALLY PROMISING PLANTS

When Khalid El Sayed published his first study on the cancer-killing potential of compounds known as cembranoids in 1998, he knew it was unlikely the discovery would be saving lives any time soon. The soft coral in which the molecule was found, *Sinularia gardineri*, grew in the Red Sea. And although it showed promise against lung, skin, and colon cancer in human cells, and leukemia in mice,[17] "it's just not easy to farm things that only grow in the ocean," he told me.

El Sayed knew, however, that cembranoids—which have a chemical structure based on a 14-carbon ring—were quite widespread in nature. "So if we could find them somewhere else where they are easier to cultivate," he said, "it could be very helpful."

What El Sayed discovered in the years that followed was surprising and ripe with potential. Yet the chemistry professor from the University of Louisiana at Monroe is still having a hard time getting funding for further research.

That's because the cembranoid source El Sayed now believes has the greatest potential for fighting cancer also causes 7 million deaths worldwide each year, mostly by way of lung disease and lung cancer. That's right: Tobacco could be a vital weapon in the fight *against* cancer.[18]

Let's let that hang there for a moment or two while we build some context: When I ask folks what plant kills more people than any other, their minds rarely go straight to tobacco. *Nicotiana tabacum* by itself, after all, isn't acutely dangerous. If you were to chew up and swallow a leaf you'd probably have one heck of a stomachache, but you're not likely to fall over and die.

But tobacco leaves are some of the most chemically loaded foliage in the world. Even before it's processed and loaded up with additional ingredients, there are at least 3,000 chemicals in tobacco,[19] and El Sayed believes the number might be closer to 5,000. Most notorious among these chemicals is nicotine, an oily liquid that acts as a stimulant when absorbed into the body's bloodstream, where it wreaks havoc on the autonomic nervous system and skeletal muscle cells, and causes one of the hardest-to-break addictions in the world.

Cigarette companies knew this for decades before they admitted it to the public, and went through extraordinary lengths to conceal the true dangers of using their products.[20] While reasonable people can disagree about how obvious it should have been that inhaling smoke of any sort wasn't a great health choice, it's worth noting that as the dangers of smoking became more well-known in the latter half of the twentieth century, cigarette company executives took the craven step of intentionally manipulating their products to make them more addictive. And while Big Tobacco was ordered by a federal court to admit to that conspiracy in 2006, the companies fought doing so for another eleven years.[21]

During that same period, US tobacco company profit margins went up more than 75 percent,[22] mostly because every time the government increases the taxes on cigarettes, the companies tack on a few more pennies, too.[23] Those profits haven't carried over to tobacco farmers, though: Market consolidation has hammered producers in North Carolina and Virginia, where the vast majority of US tobacco is grown. Nationwide, crop values fell from $1.83 billion in 2014 to $1.27 billion in 2016.

"Ultimately tobacco is a very sustainable crop," El Sayed said, noting that Native Americans were cultivating the plant long before the arrival of Christopher Columbus, and that it was first commercially cultivated in Europe in the 1550s.[24] "It's a prosperous crop, agriculturally very valuable, and many states already depend on it economically."

With recent research showing tobacco cembranoids have potential for inhibiting the growth of new blood vessels in breast and prostate tumors,[25] El Sayed figured the time was ripe for further investment in research related to beneficial uses for the plant—which could, he said, potentially save jobs and lives at the same time.

Now, if you're looking for a grant aimed at preventing smoking and tobacco use, you might be in luck: The Centers for Disease Control and Prevention, the National Center for Chronic Disease Prevention and Health Promotion, and the Office on Smoking and Health are among the US government agencies literally giving money away for science-based prevention. But if you're trying to demonstrate beneficial uses for tobacco, El Sayed said, "it's very difficult to get funding."

El Sayed said the tobacco lobby is almost certainly interested in his research, which also suggests that if cembranoids were left in the tobacco put into cigarettes—right now those compounds are scrubbed out, ostensibly to improve taste—it could make cigarettes less cancerous. "But there are several funding agencies that will not fund you if you have had any sort of tobacco funding," he said.

The National Institutes of Health, for instance, will look to see if researchers who are applying for grants have previously had tobacco industry funding. Many top-tier research institutes, like Johns Hopkins and the Mayo Clinic, outright prohibit their scientists from accepting cigarette industry support. And some top journals, including those in the *British Medical Journal* line of publications, won't even consider a submission connected to tobacco money.[26]

"I certainly understand," El Sayed said with a sigh, "but I'm not promoting tobacco smoking. We do know that some people are going to smoke it anyway, though, and we might be able to minimize the harm. And then, of course, it would be very good to make better use of tobacco to begin with as a pharmaceutical or a supplement."

As University of Manchester epidemiologist Anne Charlton pointed out in a letter for the *Journal of the Royal Society of Medicine* in 2004, the global success of cigarettes in the nineteenth and twentieth centuries essentially hijacked a plant that had, upon its initial introduction in Europe, "acquired a reputation as a panacea, to the extent of being called the 'holy herb' and 'God's remedy.' "

"I suggest we should set aside the prejudices generated by the ill-effects of tobacco smoking," Charlton concluded, "and examine the leaves systematically for substances of therapeutic value."

Researchers at Australia's La Trobe University have done just that, focusing their work not only on the leaves but the rest of the plant as well. In doing so, they have discovered a molecule called NaD1, found in the trumpet-shaped pink and white flowers of an ornamental species of tobacco, that can be used to conduct a precision strike on cancer cells while leaving the healthy surrounding cells unscathed.[27] That's caused Susan Lawler, the head of La Trobe's Department of Environmental Management and Ecology, to begin thinking wistfully about the future in ways similar to El Sayed. "Imagine fields of tobacco grown for their flowers instead of their leaves," she wrote for *The Conversation*, "leading to an outburst of health-conscious tobacco farming."[28]

Imagine indeed. For when we only see things as dangerous, the biggest danger is we might miss their true potential.

WHY THE WORLD'S MOST POISONOUS FROGS DON'T POISON THEMSELVES

The frog was a father, carrying its tadpoles on its back. And as I slowly approached it to get a closer look, I was taken by the way in which something could be so nurturing and so deadly at the same time.

"They're good papas," said my guide, Diego Gustavo Ahuanari Arujo, a member of the Cocama ethnic group who was showing me around the forests where he grew up near Colombia's Amacayacu National Park. "Once the eggs are hatched, they carry the babies to keep them safe."

I lifted my camera and got closer.

The frog was only inches from my lens when Arujo spoke up. "It is more likely to jump away from you than toward you," he said, "but maybe you should not take the chance?"

I snapped one more frame and then took a step away.

"OK," I said, "so now can I pick it up?"

"Sure," Arujo said, "if you want to be dead in the next five minutes."

"But how will we lick it then?"

Arujo was clearly tired of my kidding. "Fine," he said. "You lick the poison frog. But give me your phone so I can take a video of you dying. We will make lots of money on YouTube and my village will be rich."

We found several dart frogs that day, and more that night when we searched the forest by flashlight. "One of the first things people teach their children here is to stay away from those frogs," Arujo told me. "You have to teach them, because children love frogs."

Some adults do, too. Evolutionary biologist Rebecca Tarvin has spent a lot of time in the Colombian forests collecting and studying—and yes, even licking—some poisonous frogs. The "mild" frogs Tarvin tried tasted like sushi. "I could tell on my tongue the area that had touched the frog," she once told NPR, "and the feeling kept spreading until my mouth was kind of numb."[29]

Tarvin's tongues-on approach to science might not be for everyone, but much of her research is deadly serious. Just a milligram of poison from some of the species she *doesn't* lick, like the golden poison dart frog, is strong enough to kill ten adult humans. As such, the dart frog known as *Phyllobates terribilis* is often considered to be the most poisonous animal in the world.

So how is it dart frogs don't poison themselves? At least for some dart frogs, Tarvin has learned, the secret appears to be a single amino acid in the protein typically impacted by the deadly toxin epibatidine. A slight change in the shape of that receptor makes it impossible for epibatidine to attach itself to the protein. Essentially, the toxin just slides right off.

But there's a problem with that strategy: That same receptor is needed for the frogs' brains to work properly. So epibatidine-carrying dart frogs have evolved to have other amino acid changes—kind of like biological detours—that allow the protein to still do its job. Lots of species of dart frogs have evolved in this way, but the workaround shows up differently in different lineages—as when different highway engineers choose different ways to build a bypass.[30]

It's only because of these chemical workarounds that these frogs can even exist. But their existence alone doesn't ensure we can keep studying the hundreds of toxins they create; where they're *doing* that existing matters, too. That's because frogs are bioprospectors. They make their toxins by picking up novel compounds from the things they eat in the wild, which make those compounds from the things *they* eat, and so on. In captivity, dart frogs lose their toxicity. We can't re-create an entire rainforest ecosystem in a lab, so we can't study these toxins unless wild frogs stay wild.

Which, of course, means protecting the rainforests. And we've been really bad at that. Pollution and habitat loss have resulted in a world in which about a quarter of the dart frog species we know about are endangered—and there are likely many more that will leave this planet before we even identify them. When herpetologist Shirley Jennifer Serrano Rojas published her discovery of an orange-striped black poison dart frog called *Ameerega shihuemoy* in 2017, she lamented that without a conservation plan, the creature might be gone before it could really be studied.[31]

Amphibians have survived several global mass extinctions. But because they rely on clean water and damp habitats, and often live in areas like rainforests that are overly exploited for human needs, they've been particularly hard hit by the Holocene Extinction.[32] Making matters more perilous: It appears poisonous amphibians are naturally more likely to face extinction than their nontoxic kin in the first place, perhaps by as much as 60 percent.[33] It's not yet clear why this is, but researchers have hypothesized it could be because producing toxins is energetically costly, or that animals with chemical defenses become "strong" enough to move into habitats that are more marginally conducive to sustaining life—and thus become more vulnerable in the long term.[34]

Natural toxins, of course, have proven to be exceptionally good roadmaps for pharmaceuticals—and the deadlier the toxin, the better the potential. With each dart frog we lose, we're shutting down another of the world's best pharmacies.

That's why researchers are racing to synthesize as many dart frog toxins as possible. To that end, scientists scored a big win in 2016 when they discovered a twenty-four-step process for creating another deadly dart frog poison, batrachotoxin,[35] which interferes with bioelectric signaling, and may thus be a good research tool for understanding how nerves conduct electricity.

But it can take a very long time to synthesize a single toxin. The publication of a recipe for synthetic batrachotoxin came forty-seven years after the poison was first identified in the wild.[36]

One toxin down, thousands to go.

And that's just the ones we get from frogs.

HOW DEADLY SNAKES MADE HUMANS POSSIBLE

I know better than to be afraid of snakes.

Only a quarter of all snakes are venomous. Only a small fraction of venomous snakes carry a potent enough toxin, and enough of it, to kill a human. An even smaller number of those potential killers live in areas humans frequent. Many of those are shy and skittish, only attacking if provoked and, even then, often with venomless strikes known as "dry bites." And in most of the places where I spend my time, antivenom treatments for the most likely biters are readily available in local hospitals, albeit often abusively expensive.[37] All told, death by snakebite is an extremely rare tragedy in the United States. Not counting the zealots who use rattlesnakes in religious ceremonies,[38] the delusionals who refuse to seek medical treatment when bitten,[39] the blunderers who try to pick up rattlers they encounter in the wild,[40] and the occasional distraught individual who commits suicide by cobra (and yes, that's a real thing[41]), the number of people killed by snakes in the United States is often two or fewer each year—despite the fact venomous snakes exist in all forty-eight contiguous states and account for as many as 8,000 bites a year.

I know all of this. But when a traveling reptile zoo called Creature Encounters brought a yellow-and-white Burmese python—a completely venom-less snake named Narcissa—to a party my then-six-year-old daughter was attending, I encouraged her to go pet it, and then prayed to every god I could remember that she wouldn't ask me to hold her hand while she did so.

I'm embarrassed by this. It's irrational. But it turns out I'm in good company—or at least a lot of company. Ophidiophobia is one of the most common fears in the world, and it might be *the* most common animal phobia.[42]

There's a long-running nature-versus-nurture debate as to why so many people are so afraid of snakes, with research that appears to back both schools of thought. In 2011 researchers from Rutgers, Carnegie Mellon, and the University of Virginia published a study showing that, while humans appear to be hard-wired to recognize slithering animals more quickly than other animals, we have to learn to be afraid of snakes, either from a bad experience or, more commonly, from the reactions of people around us.[43] In 2017, though, researchers from the Max Planck Institutes in Germany and Uppsala University in Sweden came to a seemingly different conclusion. They studied the pupils of nearly fifty 6-month-old infants and saw rapid dilation—a common reaction to fear—when the babies, who presumably had no experience with snakes yet, were shown photos of various serpents, indicating fright could be innate.[44]

Whether my fear of snakes is something I learned, something I carry in the trenches of my DNA, or a combination of both, anthropologist Lynne Isbell—the same scientist who observed that giraffes often bend down to eat, rather than stretching for the leaves on the tallest branches of trees—believes snakes are responsible for making me who I am, and her who she is, and you who you are. Her Snake Detection Theory suggests primates wouldn't have evolved as we have were it not for the need to be alert to snakes—and the strongly selective evolutionary pressures of not being good at doing so.

Snakes, after all, are some of the world's most effective killers. In parts of the developing world where some of the most poisonous snakes live, including India, Indonesia, Nigeria, Pakistan, and Bangladesh, they still claim a lot of human lives. The World Health Organization estimates that more than 80,000 people die each year as a result of venomous snake bites, and that many more suffer injuries resulting in amputations and other permanent disabilities.[45] Not including other *Homo sapiens*, snakes kill more humans than any other vertebrate—and they can do it fast. In some cases, snake venom can be fatal in just minutes.

Over tens of thousands of years, Isbell believes, the threat posed to humans by snakes may have contributed to the creation of distinctively human behaviors, including the act of pointing and perhaps even the development of language. And, in the grander scheme of survival, snakes also might have applied evolutionary pressure to the physical characteristics defining many primate species.[46] Isbell has observed, for instance, that primates from places where there are lots of venomous snakes have better vision and bigger brains than the ones existing in places where there aren't as many of those sorts of predators. As evidence, she points to the happy-go-lucky lemurs of Madagascar who, as primates go, have small brains and poor vision.[47] The lemurs evolved, she notes, without having to worry about venomous serpents—there are eighty species of snake on that African island nation, but none of them are venomous.

There are countless other pressures at play in our evolution, though, so Isbell and her collaborators have taken a deep dive into the primate brain for neurological corroboration of the theory. And in the pulvinar, an area of the thalamus that provides fast visual-information processing, they seem to have found what they were looking for: a section of the primate brain that appears to "light up" selectively when monkeys see pictures of snakes—even though the monkeys, which happened to be Asian macaques, had been raised in captivity and had never seen a snake before. Shown other pictures, the same area of the macaques' brains wasn't engaged.[48]

One of Isbell's later studies showed that even a brief glimpse of the pattern of a snake—just an inch of gopher snake skin exposed between two green towels on the forest floor—was sufficient to capture the attention of wild vervet monkeys at the Mpala Research Centre on the Laikipia Plateau in central Kenya.[49] "My plan was to show the monkeys an inch of a snake's skin, then take it out another inch, and another until they recognized it as a snake," Isbell told me. "It didn't occur to me they would detect it with just an inch showing. That told me that it's not leglessness. It's not the shape of the snake. It's the scales. It has to be the scales."

That finding aligned with yet another of Isbell's studies, in which white-faced capuchin monkeys from South America were also shown to have unique responses to the sight of snake scale patterns. The capuchins showed greater antipredator behaviors when exposed to realistic-looking models of both boas and rattlesnakes than when they were presented with similar models that didn't have a snake scale pattern.[50]

These three primates, from three vastly different parts of the world, were all connected by an attentiveness to snakes that appears to come from a shared brain function specifically designed to help simians avoid serpents like the pit vipers of Asia, cobras of southeast Africa, and aspers of South America.

Do humans share this innate attentiveness? Isbell's frequent collaborator, Jan Van Strien of Erasmus University Rotterdam in the Netherlands, believes we do. Using electrophysiological tests of human subjects, he has demonstrated that, regardless of whether we consciously feel a fear of serpents, our brains respond to images of snakes with a spike in electrical activity, called an early posterior negativity, that far exceeds the levels occurring when we see images of other animals that are creepy and crawly—and widely seen as treacherous—such as slugs, turtles, spiders, or crocodiles.[51] A later study co-authored by Isbell and Van Strien showed significantly heightened early posterior negativity in response to snake skin over similarly colored lizard skin or bird plumage—which, once again, was unrelated to whether test subjects believed themselves to be afraid of snakes.[52] Bit by bit, the evidence is getting stronger: Humans and other primates are hard-wired to become quickly alert to, if not outright fear, snakes—all part of an evolutionary arms race Isbell believes likely began with constrictors (which makes me feel a little less sheepish about my terror over Narcissa the python), but which was turbo-charged as snakes evolved to inject their prey with some of the deadliest toxins in the animal kingdom.

WHY ECONOMIC INEQUITY STALLED DEVELOPMENT OF VENOM-BASED MEDICINES

While studies are beginning to suggest that snakes played an important role in the formation of our most visceral danger reflexes, we're only just beginning to study whether snakes and other venomous animals might have the power to impact our lives in other ways.

Writing for *BioEssays* in 2011, a team of biologists from Leiden University in the Netherlands called venoms "a grossly under-explored resource in pharmacological prospecting," and blamed a disparity in knowledge about reptiles—owing in part to fundamental fear—for "the neglect of thousands of species of potential medical use."[53]

Seeking to address that neglect a few years later, a consortium of European Union research organizations set out to create the largest database of toxins in history. But when its four-year Venomics Project ended, its list of venoms only included about 200 of the 150,000 venomous species in the world. And that's just the venomous ones—those that can inject toxins into their victims, usually through a bite or a sting. There are many more poisonous animals, like dart frogs, that can spread their toxins by touch alone or, in the ultimate act of revenge, kill their predators from the pit of those predators' stomachs.

One of the main reasons we don't invest more scientific effort into understanding venoms is the same reason we don't spend more on research aimed at preventing and treating malaria: The people who would most benefit are poor. Venom analysis, after all, has historically been centered on creating antivenoms—and the majority of people who need antivenoms are in developing nations.

Even when antivenoms do exist, there are "chronic gaps in antivenom supply globally that have cumulatively cost millions of lives, maimed millions more, and contributed to the burden of poverty and disenfranchisement that lingers heavily over many nations," David Williams of the Australian Venom Research Unit wrote in the *British Medical Journal* in 2015, responding to news that one of the world's largest pharmaceutical companies would no longer make the antivenom Doctors Without Borders most commonly uses to treat snakebites in Africa.

But Williams also noted that the lack of this particular antivenom was just insult on top of injury. Most Africans *already* couldn't access the care they needed in time to save life and limb. "Experts have been urging the relevant authorities to redress this denial of access to an essential medicine," he wrote, "without any meaningful response."[54]

Where the world community has failed to act, Williams wrote, a new breed of snake oil salesman has swept in. In Ghana and Chad, for example, the adoption of ineffective antivenoms resulted in snakebite fatalities rising more than 500 percent in a single year. Meanwhile, indifference has stymied attempts to get even a rudimentary view of how badly needed and readily available antivenoms are. When public health researchers from the Netherlands attempted to survey antivenom manufacturers, national health authorities, and global poison centers on the issue, they received responses from less than a quarter of the organizations they sought to survey.[55] Based on the information they were able to gather, the researchers concluded that health agencies in nations heavily impacted by snakebites had engaged in few epidemiological studies, infrequent training for health workers, and paltry development of national abatement strategies. With little international support to address the antivenom shortage, they wrote, those nations had focused health efforts in areas where they could be successful—which meant leaving many snakebite victims to their fates.

The difference in snakebite outcomes between the developed and developing world is striking and shameful. It means that even though

I'm afraid of snakes, I really have no reason to be, while people in less affluent areas of the world who have every reason to be afraid must accept death-by-snake as a part of life.

There has, however, been a recent surge in attentiveness to venoms. Not because the developed world has suddenly begun to care more, but rather because technological advancements have offered researchers the opportunity to evaluate venoms for purposes that impact rich nations, too.

The European Union's Venomics project, for instance, explicitly sought to create "innovative receptor-targeted drugs as well as novel therapeutic avenues" by focusing on "inflammation, diabetes, auto-immune diseases, obesity and allergies." And while those remedies could certainly "trickle down" to poorer nations, the participants in the project made it no secret that their primary goal was economic, not humanitarian.[56] Whatever the motive, though, the result has been the identification of hundreds of new proteins and peptides that researchers are turning into life-saving therapeutics.

Though enjoying newfound attention, applied venomics isn't a new field of science. In 1968, a research team from pharmacologist John Vane's lab at the Royal College of Surgeons demonstrated the venom of a Brazilian pit viper known as *Bothrops jararaca* could be used as an angiotensin-converting enzyme inhibitor, which causes the relaxation of blood vessels and results in lowered blood pressure. That discovery quickly led to the development of captopril, which gained FDA acceptance to be marketed in the United States in 1981 and is still used today to treat hypertension, kidney problems, and congestive heart disease.[57]

Captopril was revolutionary, but it was also a lucky break. Indigenous groups in Brazil had pointed researchers in the direction of the viper, which they had long used as a source of poison for the tips of their arrows.[58]

"The medicines of today are based upon thousands of years of knowledge," Vane noted in his acceptance speech for the Nobel Prize in medicine in 1982. Although he was being honored for his work in discovering prostaglandins, hormone-like substances that govern several important processes in the body, in his short banquet speech Vane made specific reference to the pharmaceutical promise of venom, adding "the new medicines of tomorrow will be based on the discoveries that are being made now, arising from basic research in laboratories around the world."[59]

And yet there was no rush to discover venom-based medicines in the decades that followed. The research was, at that point, just too hard. There can be thousands of toxic peptides in the venom of a single species—meaning scientists have to sort through tens of millions of chemical structures to figure out which does what. To do this, they first have to identify a venom that affects a biological process—one that drops a test mouse's blood pressure, for instance. Then they have to break that venom down into smaller and smaller component parts, looking for the single molecule, or combination of several molecules, responsible for the effect they've seen.

It wasn't until quite recently that advances in genetic sequencing and in vitro DNA reproduction made it possible for researchers to see, better and faster, the conditions in which a molecule becomes active. That allowed scientists to more easily find promising chemical candidates without the benefit of, as Vane had put it in his Nobel speech, "folklore" and "serendipity."

These days we can sort through the chemical structures of venoms with the help of supercomputers. And we can fuse together toxins from different animals to create designer molecules.[60] These advances have led to a surge in research interest in venom, and particularly in the venom of the deadliest creatures in the world, like the killers that most often show up on lists of the world's deadliest snakes.

Assigning a superlative designation to a snake's deadliness is difficult, since they hunt all sorts of different prey. Even among common prey, such as mice, some snakes kill faster, others kill more frequently, some produce more venom, some produce more toxic venom, and some—by virtue of size and appetite—just kill more. Nonetheless, there are a few snakes that show up on just about every herpetologist's you-don't-want-to-mess-with-this-one list, including the inland taipan, the black mamba, and the saw-scaled viper.[61]

The inland taipan, *Oxyuranus microlepidotus*, was the creature that sent an Australian teenage snake-wrangling celebrity named Nathan Chetcuti into a nearly deadly coma in 2017. Often cited as the snake with the deadliest "lethal dose 50" rating—that's the amount of venom it takes to kill half of the unfortunate test mice that are injected in a lab—the taipan's venom contains a component capable of forming compounds that play key roles in inflammation and stopping bleeding. It may also have a chemical that can relax the muscles responsible for the contraction of blood vessel walls.[62]

Mamba venom isn't quite as toxic as taipan venom, but mambas are a lot bigger and a lot quicker—likely the fastest venomous snake in the world. *Dendroaspis polylepis* venom works fast, too; it quickly paralyzes the snake's prey, giving the mamba time to savor its meal. That quality intrigued researchers whose analysis of mamba venom revealed the presence of a painkiller seemingly just as strong as morphine.[63] Set against a worldwide opioid addiction crisis created in large part by doctors seeking to address their patients' pain, the discovery that mamba venom is such an effective inhibitor of acid-sensing ion channels—a principal pathway for pain in humans—has sparked hopes we might soon be on the verge of creating less-addictive painkillers.

In terms of a raw human body count, there is no genus of snake deadlier to humans than the saw-scaled viper, *Echis carinatus,* one of eight related species across Africa, the Middle East, and South Asia responsible for tens of thousands of deaths each year. It's also helped doctors save a lot of lives, as the source of an antiplatelet drug called tirofiban, which prevents blood clotting during a heart attack and can be administered during surgery to treat coronary artery blockage.[64] Tirofiban is unlikely to be the last drug emanating from saw-scaled viper venom, though. In 2017 molecular biologists did a tandem mass spectrometry analysis of venom from *E. carinatus* snakes from India, learning as they did so that the viper may be something of a one-stop shop for scores of proteins also found in the venom of a variety of other snakes, plus a few novel ones that will be further examined in the years to come.[65]

Terrestrial snakes are the most studied venomous creatures in the world. But even though there are only about 600 species of venomous snakes, we're still a long way away from a full reckoning of their pharmaceutical potential.[66]

We're even further away from understanding the poisons of Poseidon.

WHY THE WORLD'S LARGEST PHARMACY MIGHT BE UNDER THE SEA

The sting wasn't bad at first.

I was diving with my wife in Los Arcos National Marine Park in Mexico when it happened. I'd spotted a manta ray and was trying to follow it from above when I swam right through a crowd of small jellyfish.

One got me. It was like a quick jab with a needle, not even enough to get me to surface in the moment. By that evening, though, I had a welt on my back the size of a 20-peso piece. It stayed there for a long time, so red, bulbous, and ugly that I wouldn't walk around shirtless for months. I felt sorry for myself only until I did an online search for photographs of other people's jelly injuries; the results were a gag-inducing collection of tangled lacerations, constellations of pustules, and dark scabs. After that I just felt lucky.

Oceanic cnidarians, including jellyfish, anemones, and corals, use harpoon-like tentacles called nematocysts to envenomate their prey and fight off predators. Some cnidarian stings are just annoying. Others can be deadly. The deadliest to humans are box jellyfish, agile swimmers with twenty-four eyes and exceptionally potent toxins, which exist in oceans around the world and like to hang out near the beaches as much as humans do. By some estimates, the box jellies claim 100 human lives a year via a fast-acting envenomation that can stop an adult's heart in minutes.[67]

The box jelly's venom holds unbridled potential. Yet as Bryan Fry, a venomologist at Australia's University of Queensland, has noted, more scientific studies are conducted on snake venom in a single year than have ever been published about jelly venom.[68] This may be in part because jelly venom is hard to get, so Fry set out to develop a technique for making it easier.

He knew ethanol could induce nematocyst discharge in other cnidarians, so he and his collaborators gathered up some box jellies of the species *Chironex fleckeri*, also known as the sea wasp. They dipped the sea wasps' tentacles in ethyl alcohol for thirty seconds, waited a day for the proteins to disperse, and then sent the liquid through a centrifuge.[69] The result was pure sea wasp venom, a handful of new proteins and peptides to study, and a process quickly adopted by researchers examining other cnidarians. Among those who have already leaned on Fry's findings are scientists studying cold-water sea anemones, Pacific sea nettles, and the enormous lion's mane jellyfish. An entire phylum of some 10,000 venomous animals has been unlocked.

Of course, cnidarians aren't the only venomous creatures of the sea. Anyone who knows the tragic story of Steve Irwin can tell you that. Irwin, who rose to prominence as the brazen and buoyant "Crocodile Hunter," had gleefully survived encounters with crocodiles and poisonous snakes; he used to joke that his worst injuries came from parrots. But it was a rather unlikely suspect, a stingray, that caused his death in 2006 while filming a documentary series that was all-too-aptly called *Ocean's Deadliest*.

Stingrays are among 1,200 species of venomous fish that, until quite recently, have been just as widely ignored as cnidarians. The most toxic ones known to science so far—and the "so far" is important here, given the vast unexplored reaches of our ocean—hail from the genus *Synanceia*: the tropical stonefish.

Stonefish, named for their rocky camouflage, secrete neurotoxins from their sharp dorsal spikes that can induce pain, swelling, hypotension, and respiratory distress in unfortunate humans who accidentally step on one.

Although we parted ways with stonefish somewhere in the neighborhood of 500 million years ago, we've still got a few things in common. For instance, a protein found in *Synanceia* called stonustoxin appears to be an ancient relative of a human protein called perforin. In humans, perforin is used to destroy cells that have been infected or have mutated in cancerous ways, but it also causes pancreatic cell destruction in type 1 diabetes patients and transplant rejection in bone marrow recipients.

Scientists have long known that perforin works, in both its positive and negative capacities, by creating a pore on the face of a cell big enough to allow toxins to enter and kill it from the inside out. In 2015, they learned *how* it does that—it begins when two subunits of stonustoxin interact, forming a crystal structure that starts pore formation. Once they realized how the pores began, scientists could start looking for chemicals that might inhibit that process from the onset.[70] If that happens, it could help the nearly 30 percent of leukemia patients whose bodies reject a bone marrow transplant.

If there's one deadly sea creature that has more pharmaceutical potential than any other,[71] it's the fearsome cone snail. Venom from members of the genus *Conus* is among the fastest-acting in the world, and these gastropods have come to be known as "cigarette snails" because, as the story goes, if one stung you, you'd only have enough time for a single smoke before your heart stopped.

The truth is that only a few of the hundreds of species of cone snails can deliver enough venom to threaten a human, but their toxic sting does work as fast as advertised—even faster if you happen to be a passing fish, which can go from merrily swimming to helplessly paralyzed in a fraction of a second. It's that sort of speed that intrigues researchers like Frank Marí, a biochemist whose cone snail "farm" at the National Institute of Standards and Technology has been at the center of dozens of discoveries involving the neurotoxic peptide the snails carry, known as conotoxin.

Marí has his milking technique down pat. He entices the snails in his lab with a dead goldfish and waits for the animals to start swinging their toothed proboscis back and forth. Then, at the last moment, Marí replaces the fish with a latex-topped vial.[72]

That's a technique that builds on a far more hilarious process developed in the 1980s by a University of Utah undergraduate named Chris Hopkins, who inflated a condom and rubbed a goldfish on it before lowering it into a tank full of venomous snails. "The sight of an inflated condom floating at the surface with a tethered snail swinging like a pendulum below it was one of those moments that should have been recorded with a camera," Hopkins' adviser, neuroscientist Baldomero Olivera, later wrote.[73]

Olivera's lab later isolated a peptide from *Conus magus*, the magical cone snail, that was the basis for ziconotide, a spinal-injected pain reliever 1,000 times stronger than morphine for patients with severe chronic pain. Researchers suspect the thousands of additional peptides in cone snail venoms may one day be used for fighting tuberculosis, cancer, nicotine addiction, Alzheimer's, Parkinson's, schizophrenia, multiple sclerosis, and diabetes.[74]

HOW KILLER SPIDERS AND FRIENDLY GOATS ARE WORKING TOGETHER TO MAKE SHOES

Bert Turnbull's 1973 treatise on the ecology of spiders for the *Annual Review of Entomology* reads like the start of a horror movie trailer.

"They are found over the entire life-supporting land masses of the world," he wrote. "Where any form of terrestrial life exists it is safe to assume there will be spiders living nearby."[75]

And wherever they are, Turnbull's research showed, they are hungry. To support their voracious appetites, they evolved to be some of the most prolific killers in the world.

By way of comparison, humans collectively kill and eat about 400 million tons of meat and fish each year. All the whales in the ocean might eat as much as 500 tons of meat. But according to estimates based on the work Turnbull did during his long career, the global spider community kills and eats as much as 800 million tons of other animals each year.[76]

For *Washington Post* data geek Christopher Ingraham, the fact that spiders eat the equivalent weight of 8,000 aircraft carriers in meat each year wasn't disturbing enough, so he did a little additional math. "The total biomass of all adult humans on Earth is estimated to be 287 million tons," he wrote, adding that even if you tacked on another 70 million tons for all the kids, "spiders could eat all of us and still be hungry."[77]

Along with snakes, spiders are some of the most feared animals in the world by humans, and yet the world would be a tremendously more frightening place if they *didn't* exist. All those hundreds of millions of tons of spider food, after all, are made up of members of the orders Diptera (that's flies), Hemiptera (cicadas, aphids, and the like), Hymenoptera (sawflies, wasps, bees, and ants), Collembola (springtails), Coleoptera (beetles), Lepidoptera (butterflies and moths), and Orthoptera (grasshoppers, locusts, and crickets).[78]

Spiders also eat a tremendous number of other spiders, meaning that killing a spider simply leaves fewer predators for the other spiders. For arachnophobes, that's one heck of a lousy zero-sum game.

Part of the reason spiders eat so much is they've been wildly successful as an evolutionary clade. There are nearly 45,000 known species of spider. And just about all of them are venomous. What biochemical treasures are lurking in all that venom? We have almost no clue at all. Researchers have evaluated only a few thousand of the likely 10 million different active molecules in the venom of the world's spiders. That nascent search, though, has already yielded molecules that may be effective for treating muscular dystrophy and chronic pain.[79]

If there's one good way to get rich men to invest in research, it's to establish a potential cure for erectile dysfunction. And that's what scientists studying the Brazilian wandering spider, *Phoneutria fera*, have done. After learning that the bite of the spider, which is found in rainforests throughout South America, can cause a persistent erection, the researchers gathered up some older rats who, like many older men, were having a bit of trouble in that department, and gave the rodents a dose of one of the spider's toxins, called PnTx2-6. Since PnTx2-6 works differently than the active molecules in the most common ED medications, it may give hope to the one-third of men who don't respond to drugs like Viagra, Levitra, and Cialis.[80]

That's not all spiders are good for, though.

Spiders wouldn't be nearly so deadly, after all, if not for one of their most emblematic characteristics—the webs most of them spin to catch their prey. Webs' filaments are stronger than steel, and that sort of strength makes spider silk an incredibly enticing substance for makers of products like ropes, nets, parachutes, and bullet-resistant materials. And since spider silk is biocompatible—it won't be rejected by the human body—the proteins that make it at once so strong and elastic are being explored as possible materials from which to make artificial ligaments, tendons, bone, and skin.

Just one problem: Owing largely to the fact that spiders are territorial and often cannibalistic, spider silk is exceptionally hard to farm. And that's why Randy Lewis doesn't work with spiders, but goats.

Goats, of course, don't spin webs. But Lewis learned that by transplanting two spider genes into the goats in his lab at Utah State University, he could get the ungulates to produce milk that contains spider silk proteins. When the milk is frozen, separated, thawed, and filtered, the result is a fine white powder that, when turned into a solution and pulled through a syringe needle, turns into a fiber that is very light and very strong—just like a spider's suspension line.[81]

Goats aren't the only organisms capable of receiving a spider gene. Lewis and others have also experimented with silkworms, alfalfa, and bacteria. And spider thread from several of these sources is starting to see its way into the market; in 2016, for instance, Adidas announced it was working on a running shoe made from synthetic spider silk that won't just be lightweight and strong, but also will biodegrade when you're done with it, bringing new meaning to the German brand's motto "Impossible Is Nothing."

WHY MAKING MOSQUITOES LESS DEADLY COULD PRESENT AN EVEN BIGGER DANGER

Anyone who has watched Bill Gates's 2009 TED Talk about fighting malaria remembers the moment the mischievous founder of Microsoft opened a glass jar filled with mosquitoes, releasing them into the auditorium and onto an unsuspecting crowd in Long Beach, California.

"Malaria is, of course, transmitted by mosquitoes," he told the audience as he unleashed the bugs. "I brought some here, so you could experience this. We'll let them roam around the auditorium a little bit. There's no reason only poor people should have this experience."[82]

If you haven't seen the talk, I highly recommend it. And at the point Gates starts to open the jar, just about 5 minutes and 10 seconds in, close your eyes and listen to the laughter.

Gates's insects weren't actually infected, and it's clear most of the people in the crowd understood that the second he popped the lid off the jar. The audience immediately laughed and applauded. Yet Gates *still* hastened to tell his fans they had nothing to be afraid of. In a room full of very privileged people, the mere possibility there might be someone in the room who believed a crazed billionaire would put people's lives at risk was too frightening to let stand for more than a few seconds.[83]

Thanks to the efforts of people like Gates—who frequently notes the global market for hair-loss cures is greater than what we spend to stop malaria,[84] and who has spent hundreds of millions of his own dollars to rectify that shameful fact—many people now at least recognize that the animal that claims the most human lives isn't an apex predator like a great white shark or a poisonous creature like a snake, but the lowly mosquito. There are, however, several caveats to the increasingly conventional belief that mosquitoes are the deadliest animals on the planet—the biggest being that mosquitoes, all by themselves, aren't deadly at all.

By itself, the worst thing a mosquito can do to you is make you itch, and it's not even really responsible for that. When a female mosquito (the males don't drink blood) sticks its needle-like mouth into a person's skin, it also releases an anticoagulant to keep it from getting stuck by clotting blood—because if there's one thing worse than having to stick your mouth into someone's skin, it's getting trapped with your mouth in someone's skin. And it's not the anticoagulants, but the histamines created by our own bodies in response to this "bug spit," that cause us to itch. Otherwise we likely wouldn't notice the effects at all.

What is actually deadly when it comes to mosquitoes are the bacteria, viruses, and parasites these insects carry, and in particular the five types of malaria-causing *Plasmodium* parasite—*P. falciparum*, *P. vivax*, *P. ovale*, *P. malariae*, and *P. knowlesi*—which enter the human bloodstream, reproduce in the liver, reenter the bloodstream, and then begin to systematically kill off blood cells. It's nasty stuff.

Of the five malaria parasites, *P. falciparum* is both the most common and the most likely to kill, likely responsible for about 215,000 human deaths annually. That's about the same number of people who were killed in the atomic bombings of Hiroshima and Nagasaki—but it happens every year. But attacking the parasite itself hasn't proven a very successful strategy for fighting malaria, especially as *P. falciparum* has become increasingly resistant to the most common antimalarial drugs, chloroquine and sulphadoxine-pyrimethamine.[85]

"The vector is the Achilles' heel of the disease," molecular biologist

Andrea Crisanti, one of the world's foremost experts on combating malaria, told *Smithsonian* magazine in 2016.[86] If you attack the pathogen, Crisanti continued, "all you are doing is generating resistance."

But if you destroy the carrier? Game over.

There are about 3,500 species of mosquitoes. Only about 100 spread disease. And just eight—those belonging to the *Anopheles gambiae* complex of morphologically identical but reproductively isolated species—are responsible for transmitting the vast majority of malaria cases worldwide.

An increasing number of scientists have come to believe *A. gambiae* plays no meaningful ecological role in our world, other than spreading disease. They've looked. They've looked hard. And no one can find a niche this mosquito fills that, were it to simply disappear, wouldn't be filled by another creature. We've driven a lot of animals into extinction through hunting, habitat loss, and human-caused climate change, but some folks are now suggesting these particular blood-suckers, and a handful of other mosquitos that carry deadly diseases, could and should be the first animals humans intentionally eliminate from the face of our planet.

The first time I heard of the idea of intentional extinction was in a 2010 article for *Nature* by journalist Janet Fang. "Ultimately," Fang wrote, "there seem to be few things that mosquitoes do that other organisms can't do just as well—except perhaps for one. They are lethally efficient at sucking blood from one individual and mainlining it into another."[87]

Fang did find some cautious resistance to such an idea. But not a lot. It turns out that even bioethicists hate mosquitoes. And even if they didn't, given that we live in a world in which humans are already driving animals into extinction by the thousands, it's hard to make an ethical case that one species of mosquito is worth millions of human lives.

Conspicuous by its absence in Fang's article was the fact that, at the time, some people were already at work making the world's first intentional extinction happen. Indeed, by the time Fang was asking whether it should be done, a group funded by Gates had already taken big steps toward that end on the Caribbean island of Grand Cayman. The Cayman study, which flew under the international radar for more than a year, was intended to see whether the release of male mosquitoes, genetically engineered to be sterile, could reduce the prevalence of dengue fever. And that's just what happened, according to Oxitec, the UK-based company that ran the trial. Over the course of six months, the company had released more than 3 million sterile males, overwhelming the population of natural, fertile males. That resulted in an 80 percent reduction in the mosquito population—enough to effectively wipe out dengue in a town of around 3,000 residents.[88]

For years scientists had been debating how to carry out such experiments—and the World Health Organization was in the process of drafting guidelines for how such a study should be undertaken.[89] But Oxitec hadn't waited for any of that. It argued there was no need to get approval from anyone other than the local government. And with its success as the backdrop for the announcement that it had released genetically modified insects into the wild without informing the international community, the criticism over the company's actions was rather muted.

And Grand Cayman is, after all, an island. Chances were quite slim that the Oxitec mosquitoes—whose sterility ensured their impact on their own species could only last a generation—would have any effect on the larger world.

It's for those very same reasons, though, that it wouldn't be easy to scale up a mosquito eradication program like the one Oxitec first embarked upon in the Caribbean. That process might kill off a lot of mosquitoes in a generation, but not all of them.[90] To kill all the skeeters, the sterile males would have to be released again and again, across the entirety of a species' habitat, for each and every mosquito species targeted for elimination.

That's where gene drive comes in. Instead of pushing a species into extinction, some scientists have now decided, the better solution would be to fundamentally alter it so that it could survive, but in a less dangerous form. In 2015, a team from the University of California at Irvine used the gene-editing-made-easy technique known as CRISPR-Cas9 to genetically engineer mosquitoes that carry *P. faciparum*–killing antibodies. They further rigged the system so the antibody gene would be supremely dominant, at a passdown rate upwards of 99 percent,[91] ensuring just about every offspring—and their offspring, and their offspring—would be incapable of spreading malaria. Given that the cycle between birth and breeding in mosquitoes takes a matter of days, it wouldn't take long for such a gene to overwhelm a broad population.

You might think the opposition to completely pushing a species into extinction would be greater than the opposition to changing it, ever so slightly, to make it less dangerous. As it turns out, though, that hasn't been the case. Whereas a lot of bioethicists seemed rather "meh" on intentional extinction, the opposition to releasing gene-drive-altered insects into the wild has been loud and clear. That's because even if we were to begin such an experiment on an island, it would be difficult, if not ultimately impossible, to keep it there. Gene-drive modification is like the Energizer Bunny: Once unleashed, it just keeps going and going and going, until an entire species has been wiped out or turned into something else.

To Eleonore Pauwels, one of the world's most innovative thinkers on the subject of creating "actionable" ethical standards for health and genomic technologies, the dangers are clear. "We now have the power to hijack evolution," she said in 2016.[92]

To University of Oxford philosopher Jonathan Pugh, the real danger in "playing God"—by deciding not just whether an animal lives but how it lives—is that "humans are not omniscient and we might overlook the possibility of devastating unintended and unforeseen consequences."[93] And the possibilities that we *haven't* overlooked are scary enough.

In 2016, James Clapper, who was then the US director of national intelligence, warned the Senate Armed Services Committee of "far-reaching" national security implications[94]—bioweapons the likes of which were nothing more than a dystopian backstory until just a few years ago. If you can change *A. gambiae* so it can no longer transmit malaria, after all, there's not much stopping you from changing it—and thousands of other mosquito species—to carry all sorts of other deadly diseases.

What's to stop a would-be terrorist, one with a bit of computer science and biology savvy,[95] from using a mail-order CRISPR gene-editing lab to do just that? The scary answer is: Not much.[96]

And even if those with nefarious aims stay away from the bug-making business, what is to prevent organizations like Oxitec from releasing gene-drive-altered animals into the world, risking unintended consequences? The even scarier answer is: Almost nothing.

Under current international law, each country has a right to choose for itself what biological risks it exposes itself to. Recognizing creatures like mosquitoes don't pay much heed to international borders, however, many nations have signed on to the Cartagena Protocol on Biosafety, which governs transboundary movement, transit, and handling, and the use of "living modified organisms."[97]

But it's no coincidence that when scientists sought to release *Wolbachia*-infected mosquitoes in an attempt to control dengue fever and chikungunya (the *Wolbachia* bacterium kills off offspring during embryonic development) they chose Queensland, Australia, to do it. Australia wasn't a signatory to the protocol.

As any fan of the TV series *Dexter* can attest, it's awfully hard to root against a killer of killers. Is it really so bad for scientists to go "venue shopping" when all they're trying to do is find countries where there are fewer hurdles to wiping out deadly diseases?

Maybe not. But how deadly does something have to be in order for us to forgive a little ethical flexibility?

If we're willing to overlook the overlooking of international standards when it comes to mosquitoes, what about freshwater snails, which transmit parasitic flatworms that cause schistosomiasis, resulting in an estimated 200,000 deaths per year, according to the World Health Organization? How about snakes, which kill 100,000 or more people each year? And what of dogs, who are responsible for up to 60,000 human deaths per year, mostly through the transmission of rabies?

How many organisms are we willing to change, at a genetic level, to protect ourselves? How will we decide which genes are best suited for the task? And who gets to choose?

How we answer these questions when it comes to the world's deadliest organisms will directly inform our relationship with Mother Nature for the rest of history.

Soon we'll have to choose. I sure hope we choose wisely.

Chapter VIII

SMARTER ALL THE TIME

Why the Most Intelligent Life-Forms Ain't Us

Tanner couldn't see. And that didn't matter one bit.

The sixteen-year-old Atlantic bottlenose dolphin was wearing foam cups over his eyes, but every time I gave his research partner Rainbow a silent hand signal, Tanner waited for a beat and then did exactly what the older porpoise was doing.

I'm not sure I would have been so impressed if the dolphin duo had been underwater. Just about everyone knows what good echolocators dolphins are; it's often said they can tell the difference between a ping-pong ball and a golf ball from a hundred yards.

But Tanner wasn't putting his head underwater, so he couldn't have been echolocating. "He's using passive hearing," Dolphin Research Center trainer Emily Guarino whispered. "We're pretty sure that he's honing in on the sounds of splashing, and figuring it out from there."

One of the foam cups fell from Tanner's right eye, and the dolphin flipped it over, balanced it on his nose, and dropped it back into Guarino's hand. As she reaffixed the blinder, her assistant, Christina McMullen, showed me the next hand signal to use—a wave intended to elicit a response in kind.

I must have waved a little overenthusiastically, though, because instead of waving back at me, Rainbow started turning in circles. And Tanner, not knowing I'd done anything wrong, simply followed his friend. Tanner was better at imitating Rainbow, it seemed, than I was at imitating McMullen.

Tanner can imitate humans, too. But at first he struggled at it. Researchers from the center, where ninety staff members care for about thirty dolphins in the Florida Keys, figured that was because humans—with their awkward arms and wobbly legs—make different sounds in the water than dolphins. Tanner, they surmised, was having a hard time creating a mental map of what his human trainers were doing.

But after grappling with the challenge for a bit, Tanner did something that surprised them: He briefly dipped his head into the water and, when he emerged, did exactly what the trainer was doing—bobbing up and down, spinning around, or turning upside down.

"When passive hearing wasn't enough, he switched to echolocation," Guarino said. "He was problem-solving."

To Guarino, this was an "aha!" moment. Dolphins don't just process sound in ways that humans can't, she said. They can recognize when one sensory strategy isn't working and use another, too—also in ways that humans can't.

These aren't the only ways dolphins are our cognitive betters. They also appear to be exceptionally self-aware, showing self-directed behavior in front of a mirror as early as seven months of age. Humans generally don't demonstrate the ability to recognize themselves in a mirror until they are at least a year old. Chimps, which have a similar lifespan to dolphins, don't demonstrate the ability to differentiate between their own reflection and another ape until an even later age.

Animal behavior and cognition expert Rachel Morrison was awestruck the first time she saw a young dolphin named Bailey identify herself in an underwater mirror. She was working that day with the legendary dolphin researcher Diana Reiss, recording from behind a one-way mirror at the National Aquarium in Baltimore.

In the video, Bailey swims up, nods her head, and brings her left eye up close to the mirror.[1] And for people who spend hours on end, for years upon years, watching porpoises interact with one another, that was clearly a behavior that was different from the way dolphins act around other dolphins.

"I remember saying, 'Are we seeing the same thing?' " Morrison recalled of the moment she shared with Reiss. "We always try to be super conservative, and Bailey had actually shown some self-directed behaviors even earlier, but we wanted to be sure—and this was so clear."

Later videos show Bailey watching herself blowing bubbles, bumping her blowhole into the mirror, and aligning her body to be able to see a mark that researchers had drawn on her chest with a pen. Even to my untrained eyes, it sure seemed like she was admiring and studying herself in the mirror.

Morrison said it can be contextually helpful to understand that dolphins seem to be able to recognize themselves in a mirror earlier than we can, but she's often worried people might get the wrong idea about what the research shows.

"We're certainly not saying that dolphins are more intelligent than humans are," she said. "What we are saying is that it's interesting to look at how precocious they are in this part of their development."

Cognitive scientist Kelly Jaakkola agrees. It's important not to engage in apples-to-apples comparisons, the Dolphin Research Center research director told me. After all, there are many ways in which, were they human, dolphins wouldn't seem so smart at all. Dolphins have a really tough time, for instance, with object permanence—the understanding that things that have been observed once will continue to exist even when those things aren't being observed.

To demonstrate this, the center's trainers introduced me to Jax, who had been taken from the St. John's River near Jacksonville after being badly injured as a calf. Jax may be missing parts of his dorsal and fluke, possibly as a result of a run-in with a bull shark, but he's a strong swimmer and, by nearly every trainer's account, one of the cleverest dolphins at the center.

But when one of Jax's favorite toys, a stuffed crocodile, was put into a bucket, and the bucket was moved a few feet, Jax struggled to find the fuzzy croc. And when the toy was put inside a small bucket, then transferred into a larger bucket, he struggled even more.[2]

To us, Jaakkola said, a slow and simple "shell game" like the one we played with Jax, "would be a no-brainer, but to them it's confounding."

But that doesn't say anything at all about dolphin intelligence, Jaakkola said. "It really just says something about us."

Tursiops truncatus isn't smart *like* us, she said; it's smart in completely different ways. And to that end, Jaakkola said, research demonstrating that dolphins struggle to understand something in human ways is every bit as important as discoveries that show dolphins can think in ways that are advanced in comparison to people. "If we only look at areas that we think of as great strengths," she said, "we're really missing a big part of the picture."

And that picture is one in which dolphins and humans don't exist on different parts of the same cognitive scale, but rather on very different spectrums that, in the 95 million years that separate us from our last common ancestor, have vastly diverged as well.

"Intelligence isn't one thing that you have more of or less of," Jaakkola told me. "It's like comparing Mozart and da Vinci and saying, 'OK, now who was more talented?' Both were geniuses, but they were working in very different mediums."

And we should be thankful for that. Because if animals all thought like we do, we'd probably be in a lot of trouble.

WHY DOLPHINS DON'T KILL US (WHEN CLEARLY THEY SHOULD)

It was one of the hottest tickets at the 2009 Sundance Film Festival, and I felt fortunate to have gotten a spot in the theater, even though the only remaining seat was on the far left side of the front row. Like many others in the audience that afternoon, I'd heard *The Cove*, a documentary about the annual dolphin slaughter in Taiji, Japan, could be hard to watch. And it was. The most violent part of the film—secretly obtained video footage and underwater audio of the slaughter—only lasts a few minutes, but the scenes of fishermen driving spears into these terrified animals, as the water around them turns from dark gray to bright red, were enough to leave a lasting, and horrifying, impression.

In the weeks after I watched the film, I couldn't take my mind off one particular scene in which several men, wearing wetsuits and snorkels, stood chest-deep in the water of the cove amidst dozens of dolphins, calmly maneuvering the animals toward a chute where they would be killed. The dolphins, clearly distressed, were thrashing about and trying to flee.

But there was something that, inexplicably, the dolphins weren't doing: They weren't attacking the divers.

It's not like they couldn't have. A small adult bottlenose dolphin is bigger than most large humans. A large dolphin can weigh as much as seven men, and one study estimated their force output is stronger than ten Olympic athletes.[3] Their teeth are razor sharp. Dolphins have been known to attack and kill some species of shark.[4]

They could have killed those men. But they didn't.

Laura Bridgeman, who has witnessed the Taiji slaughter, was also baffled by this behavior. "These dolphins are going—unwillingly, yet relatively peacefully—beneath a tarpaulin-covered area where they will be restrained by their tails beside one another. A metal spike will then be driven slowly into their heads," Bridgeman wrote in 2013. "Their family and friends are thrashing about in the water all around them, utterly exhausted and likely terrified . . . What species of wild animal do we know of—or even who among our own—would not lash out when backed into such a corner?"[5]

That might seem to be a particularly perplexing question in light of research showing that cetaceans have tremendously large limbic systems, the "emotional" part of the brain. It is much bigger than our limbic system, but even among other whales and porpoises, the dolphin's limbic system is quite spectacular. It has nearly twice the neuron density, for instance, as the humpback whale.[6] Shouldn't dolphins be even more "emotional" than we are?

Yes. And also no.

Dolphins may indeed have a heightened emotional existence, but they also might be able to *control* their emotions to a far greater degree than we can. Because of the way *T. truncatus*'s limbic system is spread out across its brain, some researchers believe dolphins might not have emotional and rational "sides" of their brains that work in conflict with each other (as the road-raging human brain does all too often). Rather, Wild Dolphin Project research director Denise Herzing believes emotion could be a much more fully integrated part of every dolphin thought and action. That might result in even deeper emotional attachments to one another than humans have, and perhaps even a greater sense of empathy—or a cetacean analog of that human emotion—for other creatures.

What's more, research stemming from humans who have suffered brain damage indicates that the ratio of neurons in the limbic system to neurons in the neocortex, which is associated with the processing of sensory information such as sight and hearing, impacts the ability to exercise emotional self-control. Dolphins have a much higher ratio than humans. Thomas White, a scientific adviser to the Wild Dolphin Project, believes that may explain "why dolphins don't act more aggressively to humans in circumstances that would probably provoke a violent response in humans."[7] In other words, if there is something that a dolphin won't do in typical circumstances—like taking the life of a human—a heightened emotional state might not change that fact.[8]

If that's true, it would be a tremendous indicator of emotional intelligence.

For far too long, humans have relied on intellectual reasoning as the key measure of intelligence. Recent decades have given us a better understanding of the vital importance of emotional intelligence—the capacity to be aware of and control one's emotions. The greatest tests of emotional intelligence, of course, come when we are faced with great stress. And we fail these tests with such regularity that low emotional intelligence might as well be called part of being a human.

But part of being a dolphin, it appears, is being in control at all times. Indeed, dolphins are even conscious breathers—they must think about every breath they take, planning and executing each act of respiration while negotiating all of the other aspects of their lives.

But is it actually intelligent to accept slaughter over fighting back?

That might depend on what dolphins know about us. Do they recognize, for instance, that we humans have a particularly hard time controlling our emotions? That we have a history of responding with devastating brutality to even the smallest perceived assault? That we are, by far, the creatures on this planet who are most prone to sadistically torture other animals, and in ways that make the most brutal of animal-on-animal encounters look like absolute mercy? Do they recognize that not only would attacking a human make things worse for themselves, as individuals, but to the community of other dolphins to which they are so deeply attached?

If so, then they may have a lot in common with elephants.

HOW ELEPHANT MEMORIES CAN HELP US UNDERSTAND TRAUMA

It may not quite be true that elephants never forget, but researchers believe that old adage probably isn't far off.

The world's largest land mammal has the biggest brain of any animal on terra firma, and the largest temporal lobe, which, among other things, helps process short-term information into long-term memories.

Elephants can, for instance, remember safe routes to food and watering spots decades after last visiting those locations. There's also very good evidence that elephants recognize other elephants they've met, even decades later. At an elephant sanctuary in Tennessee in 1999, for instance, an Asian elephant named Jenny met a newcomer named

Shirley. The two stomped about, bellowed, and explored one another with their trunks with such excitement that the park's workers suspected they must have known each other. Indeed they did. The caretakers later learned that Jenny and Shirley had crossed paths before as part of a traveling circus. The crazy part? Their previous meeting had lasted just a matter of weeks, nearly a quarter-century earlier.[9]

Consider those facts in a human context: Could you remember how to get to a restaurant you visited only once, decades ago, without the help of the maps app on your phone? Would you remember someone you briefly crossed paths with twenty-five years ago without checking Facebook?

What factors make something or someone memorable in that way? Given the remarkable ways in which they remember, elephants may have the capacity to help us understand why our memories work the way they do—particularly in relationship to past trauma.

Even within our own species, we still know very little about the long-term effects of physical, mental, and emotional distress. Plenty of people have dismissed the idea that trauma has the power to "rewire" our brains. Dinesh Bhugra, the former president of Great Britain's Royal College of Psychiatrists, has gone so far as to argue that post-traumatic stress disorder is not a mental health condition but rather a cultural one. He's even suggested the mere act of offering people a label "inevitably" causes those people to behave according to the label.[10] Simply put: Bhugra is suggesting that unintentional social brainwashing is to blame for an epidemic of PTSD.

Most other psychiatrists have come to different conclusions, but bold psychological hypotheses are easily spread and hard to disprove; I've heard Bhugra's claim echoed by commanding officers and Veterans Administration officials in the United States. "If you journalists just stopped writing about post-traumatic stress," a VA official once told me when I asked about backlogs in the new patient intake process, "we'd have a lot fewer people in here thinking they need help."

How do you fight back against ideas like that? I didn't know it back then, but elephants are a key.

We're not the only creatures who suffer, and it's increasingly clear we're not the only creatures who hold onto our pain for a long time.

That's what Les Schobert helped me see when we met in the fall of 2008. The former curator of the Los Angeles Zoo was dying of lung cancer. Before he left this life, it seemed, he wanted to give people an honest understanding of what they were seeing when they went to a circus or a zoo—and he spoke with grave candor about an industry built around locking up creatures that were born free.

Unlike other animals we've subjugated for riding, working, or companionship, Schobert told me, elephants have never been truly domesticated. It is exceedingly difficult to breed elephants in captivity, and the gamble one takes in trying to do so is not small. The long history of human "care" for elephants, therefore, is one that begins in the vast majority of cases with a calf, usually a perfectly healthy one, being taken from the wild.[11] And, as you might imagine, that's not an easy a thing to do.

"The only way you could do it was to kill the mother first," Schobert said a few weeks before he died. "You couldn't get a baby elephant away from its mom in any other way. You had to shoot the mom and then collect the babies."[12]

You might think that elephants on wildlife reserves would have been spared such barbarity. But that, in fact, is often where it occurred.

While the population of elephants in South Africa fell to fewer than 200 earlier in the twentieth century, by the 1980s it had steadily rebounded to more than 8,000, pushing against the capacity of the country's wildlife parks and its financial means. To alleviate the strain, the South African government did something that, given the plight of African elephants nearly everywhere else on the continent, might seem absurd: It approved the killing of more than 3,000 elephants and permitted the capture of juveniles orphaned in the cull. Hundreds of young elephants captured in this way were sold to zoos around the world.

A few years ago, Professor Karen McComb, who studies animal behavior and cognition at the University of Sussex, began a discussion with colleagues around the world about what happened to the elephants who survived the cull in South Africa. Did they recall the trauma? Nobody really knew.

McComb and her colleagues set out for Pilanesberg Park in South Africa, where a few herds made up of elephants that had survived the cull had been transplanted, and Amboseli National Park in Kenya, where fortunate herds of elephants had lived their lives relatively unperturbed by humans. Then, using custom-built loudspeakers that would make even the most discerning audiophiles drool on their subwoofers, the researchers blasted the herds with a variety of recorded elephant calls. The first set of calls were designed to be non-aggressive: recordings from younger members of the elephants' own herds. The second set of calls were intended to sound more threatening: recordings from older members from unfamiliar herds.

"The differences in the ways the herds behaved were really quite profound," McComb said.

The elephants who had lived relatively peaceful lives reacted to the threatening sounds by perking up their ears, raising their trunks to sniff the wind, and bunching up together. The elephants who had survived the cull didn't react any differently to the threatening calls than they did to the non-threatening ones. They were, it seemed, numb to the emotions they should have been feeling in those moments—not unlike the way in which people who have past traumas can struggle to align their emotional responses with their physical reality.

No one, of course, had *told* these elephants about post-traumatic stress disorder, but the research team concluded that what they were witnessing was, at least in part, the long-term effect of the psychological distress suffered by the Pilanesberg herds—another brick in a wall of evidence demonstrating that post-traumatic stress in humans isn't just a cultural construct, but one that was embedded in our genes, and in the genes of creatures with which we share our ancestry, tens of millions of years ago.

Conservationist Ee Phin Wong is hoping to build a better understanding of what causes a traumatic experience to become a lifetime debilitation by studying Asian elephant stress hormones. "There have been many studies in humans that show genetic and epigenetic changes in the glucocorticoid system of people who have experienced traumatic stress," she told me. "So I wanted to see if this was true for elephants, too."

To do that, Wong followed a wild Asian elephant that was showing peculiar behaviors along a roadway in Malaysia. The elephant had been relocated from another area, and Wong suspected the animal might be hormonally imbalanced as a result of that experience. Sure enough, when the elephant's dung was collected and processed, the test revealed low levels of glucocorticoids. At a hormonal level, elephants and humans respond to traumatic stress in similar ways—ways that remain long after the trauma has passed.

The Amboseli and Pilanesberg experiments, and Wong's studies in Malaysia, have only scratched the surface of what we might learn about elephants and trauma, in ways that may help us better understand humans and trauma. Since the vast majority of captive elephants in the United States are wild born and have suffered some trauma, if not a significant degree of it, during their capture and relocation, they're rife with potential for research that we simply could not conduct otherwise.

When scientists perform trauma experiments on mice, they do so by traumatizing the mice—for instance, by removing pups from their mothers, shocking them with electricity, putting them in cages next to hungry predators, or striking them in the head with a metal rod. For better or worse, we tolerate this sort of treatment on rodents but, quite subjectively, would not if it were done to animals like elephants with which we have a greater affinity.

Yet elephants clearly have something to tell us about how evolutionarily successful, large-brained, long-lived social mammals with good memories respond to trauma. And that may be especially true when it comes to epigenetics—the study of how genes are expressed through chemical processes like methylation, and how changes in expression can be passed down from one generation to the next. Trauma, as it turns out, is a powerful influencer of genetic expression. In one study demonstrating this effect, researchers sprayed the cages of laboratory mice with acetophenone, an organic compound often used in fragrances resembling cherries, strawberries, and honeysuckle, then subjected the mice to electric shocks. As you might expect, the rodents quickly came to associate the smell with danger, exhibiting signs of stress whenever it was sprayed on their cages. The researchers then used male mice who had been subjected to this experiment to create a second generation of mice, not by natural breeding but through in vitro fertilization. This prohibited any social transmission of the dangers of the acetophenone smell; the mice fathers couldn't tell their offspring that a whiff of honeysuckle meant impending pain. Nonetheless, the next generation also exhibited signs of stress in the presence of that smell, as did yet another generation after that.[13]

Epigenetic inheritance is relatively easy to study in animals like mice, which become sexually mature at six to eight weeks old. It's a much different story for long-lived animals. Writing for the journal *Biology* in 2016, University of North Texas professor Warren Burggren lamented that it's no problem "to do thought experiments in epigenetic inheritance, but quite another thing to carry out actual experiments," owing to the costs in time, money, and space required. It's not surprising, Burggren noted, that when studying epigenetics, scientists tend to stick to animals like fruit flies and worms. "Put simply," he wrote, "one does not leap to study epigenetic inheritance in elephants, with their life span of up to 70 years."[14]

To which we should probably start asking: Why not? While it would be difficult, if not impossible, to morally and economically justify intentionally traumatizing and then maintaining a study-sized population of elephants over decades solely for the purpose of studying epigenetic expression and inheritance, what about the hundreds of elephants we're already keeping?

Take, for instance, the young elephant I often visit at my local zoo, Zuri. Her mother, Christie, was torn from the wild in South Africa's Kruger National Park and transported to the United States when she was less than a year old. What might be revealed by a study of the methylation of Zuri's genes? What did she inherit as a result of Christie's trauma?

As we ask more questions like these, we're finding that concepts like empathy and trauma aren't only human constructs, or even unique human experiences, but rather a part of our genetic heritage going back a very long time—evolutionary muscles that have been flexed trillions upon trillions of times.

Elephants seem to have developed similar muscles. Maybe stronger ones. Maybe even the strongest ones. Like dolphins, they may be some of the most emotionally intelligent creatures in the world, and their ability to remember is clearly quite spectacular, if not indeed superlative—and that means we have a lot to learn from those tremendous beasts.

So far, we've only discussed animals that are our close evolutionary siblings—those whose intelligence evolved down a very similar path as ours and didn't diverge until a relatively short time ago.

There's lots to be learned, though, from creatures who took a very different evolutionary path toward getting an intellectual leg up on the rest of the world.

HOW OCTOPUSES ARE LIKE A FOOTBALL TEAM

His name was Inky the octopus, and in 2016 he became a worldwide sensation when he moved the lid off his tank at New Zealand's National Aquarium, crawled down the side of the container, located a drainpipe, and slid his way to freedom in the Pacific Ocean.

I was surprised when I first heard about Inky's epic escape. The cephalopod's keepers weren't. It turns out that octopus escapes are a rather common occurrence. Octopuses[15] are widely regarded to be the smartest invertebrates in the world, and since invertebrates make up about 97 percent of all the world's animals, being the smartest is no small deal.

In the ocean, octopuses have been known to break into crab and lobster traps and steal the catch, to stow away on fishing vessels for a crustacean buffet, and to pick up and carry coconut shells for use as shelters at a later time. At least one species, *Thaumoctopus mimicus,* is capable of changing its color, skin texture, and body configuration to imitate a variety of other sea creatures, including flounders, jellyfish, sea snakes, and sponges—apparently deciding which animal to imitate based on what predators or prey is nearby.

In captivity, they've been documented learning to pull a release lever for food, navigating mazes, raiding other aquarium tanks for captive prey, and even studying other octopuses for clues about how to solve puzzles.

And octopus intelligence develops much faster than human intelligence does. It has to; most species of octopus die before they reach two or three years of age. Philosopher Peter Godfrey-Smith, the author of *Other Minds: The Octopus, the Sea, and the Deep Origins of Consciousness*, strongly suspects that octopuses' short lives and big brains are evolutionarily connected. Early cephalopods carried shells just like their fellow mollusks, snails and nautiluses. Losing the shell came with some distinct evolutionary advantages—the ability to squeeze their boneless bodies into incredibly small spaces to hide from predators and hunt for prey, for instance. But it also came at a pretty big loss by making them a lot more vulnerable to their other predators. That, Godfrey-Smith believes, put a premium on intelligence. The smarter an individual octopus could be—and the faster it could develop that intelligence within its lifespan—the more likely it would be to survive.[16]

This set of evolutionary trade-offs gave rise to a sort of intelligence that looks in many ways like ours, but that evolved some 400 million years before our species even came along—and more than 100 million years after our lineages parted ways. In a more chronological manner of thinking, octopuses didn't evolve to think like humans do—humans evolved to think like octopuses do. Octopuses, biologist Sydney Brenner has proffered, "were the first intelligent beings on the planet."[17]

As the winner of pretty much every prestigious award that exists for biological discoveries, including a Nobel Prize, Brenner could have spent the final years of his career focused on absolutely anything. He once told a fellow scientist, though, that he wanted to be "like an octopus with tentacles everywhere" who knows "everything about everything."[18] And, as it happens, one of those tentacles fell upon cephalopods. It was Brenner who initiated an effort to sequence the first octopus genome—an effort that was completed in 2015.[19]

What resulted from that effort was the discovery of hundreds of genes that had never been seen before in any other animal—many of which appear to play a role in the development of the octopus's unique central brain and the large groups of neurons in its tentacles, each of which enjoys significant neural autonomy.

An octopus operates a little like an American football offense. The quarterback might call a play, but the other players must make independent decisions as they seek to carry out those orders. Similarly, an octopus's central brain might tell one or all of its tentacles to go explore a space under a rock for food, but once the tentacle has been dispatched, it can act independently to explore the spaces and crevasses around that stone, or to grab onto whatever tasty crustacean it finds there.

The unique genes that make this sort of intelligence possible have offered us new insights into convergent evolution, the biological phenomenon in which similar features show up in species of different lineages—and sometimes in species of vastly different lineages. Often this occurs because genes that have been conserved, sometimes over very long periods of time, have maintained shared potential, like the aerobic energy-metabolism genes that helped address the needs of large-brained humans and even larger-brained elephants, long after our two species diverged. Sometimes, however, animals evolve to meet similar challenges in very different ways.

Millions of years of evolutionary pressures on octopuses worked together to design problem-solving neural networks through distributed, "football team" intelligence. And that has given scientists a new model for solving a key challenge in the brave new world of drone ubiquity.[20] The idea is that networks of small drones—familiar now to anyone who watched Intel's 1,200-drone light show at the opening ceremony at the 2018 Olympic Games in PyeongChang—could both take orders from a single source and also problem-solve on their own, both freeing up centralized processing power and providing redundancy when weather or geography interferes with connectivity.

Octopus-like artificial intelligence could change the game for drone-based search and rescue—which has already changed the game for search and rescue in general. Since 2013, a group of mostly recreational drone enthusiasts called Search with Aerial RC Multirotor, or SWARM, has donated its eyes in the sky for hundreds of aerial search-and-rescue operations in support of public agencies. By 2018, the group had become the largest all-volunteer search and rescue organization in the world. Most drone operators, though, can only control one aircraft at a time; distributed networks of semi-autonomous drones could fan out over much larger areas in much shorter amounts of time.

Octopuses are just one of many very smart organisms, though, that are being used as models for how we design and use artificially intelligent technologies.

WHY SINGLE-CELLED ORGANISMS ARE KEY TO BUILDING BETTER ARTIFICIAL INTELLIGENCE

I adore the Ig Nobel Prizes. Not because the annual awards, which celebrate seemingly trivial achievements in research, are funny—although they can be—but because they're often awarded to an experiment that most scientists wouldn't even consider doing, precisely because it looks so trivial.

A lot of scientists are in on the irony now. That's why honorees, fully aware that they are about to be lampooned, are almost always present for the awards ceremony, a variety show of sorts held at Harvard University, and the awardee lectures, hosted down the road at the Massachusetts Institute of Technology. Those who can't show up in person often send in humorous videos, like when the discoverers of a cave-dwelling insect that has a male vagina and a female penis delivered their acceptance speech from inside a dark cavern.[21]

When Toshiyuki Nakagaki, Ryo Kobayashi, and Atsushi Tero from Hiroshima University in Japan won an Ig Nobel in 2008, they didn't miss the chance to visit Cambridge, Massachusetts, to celebrate their discovery that slime molds can solve puzzles. After singing a three-note song—this was among the briefest performances of the awards ceremony—Nakagaki offered a nearly-as-short defense of the tiny organisms he's spent much of his career studying. If you want to insult someone in Japanese, apparently, you might call them "a single cell, which means 'almost stupid.' But now is the time to say 'Objection!' " he declared. "The single cell organism is much smarter than we usually thought!"

How much smarter? Maybe smarter than us. In concert with the other cells in our brains, our neurons can certainly do some amazing things. But human brain cells are transmitters that, in order to solve even the simplest of problems, must work in concert with tens of thousands of other brains cells. Single-celled slime molds, it turns out, can do that all by themselves.

At first Nakagaki didn't know how *Physarum polycephalum*, a large amoeba-like cell, was able to identify the shortest path between two exits in a maze. But he suspected what he was seeing might be a primitive form of intelligence, one that might exist in other single-celled organisms as well.[22]

In the years following his team's discovery, other scientists added more evidence to that hypothesis. At Princeton, researchers watched amoebas from the genera *Dictyostelium* and *Polysphondylium* for ten hours at a time, recording every turn the little creatures took. In the process, they discovered that once an amoeba has turned one way, it is twice as likely to go the other direction the next time it turns—which indicates the single-celled organism has something of a memory, helping them avoid moving in circles and thus optimizing foraging routes.[23]

And it's a long memory, at that. Individual amoeba live for a few days on average, and only make a turn every few minutes. The human equivalent would be remembering something you did twenty days ago. On most days, I can't even remember what I was wearing the day before.

Nakagaki's team later demonstrated that if they heated up and cooled down *P. polycephalum* in regular sixty-minute intervals, the slime mold would slow down its speed in anticipation of the cold, demonstrating not just the ability to remember and anticipate a pattern but also a remarkable sense of timing.[24]

But how? One potential answer is in the cell's cytoplasm. As a cell moves, tiny particles move through its intracellular fluid, creating channels. As long as the cell continues to move in ways consistent with how those grooves were created, the channels are reinforced, creating a "memory." If the environment changes, though, and the cell has to react differently, the channel breaks down and the memory is erased.

Wanting to mimic that basic memory storage device, University of California at San Diego theoretical physicist Massimiliano Di Ventra designed a simple circuit that employs a memristor—an electronic element that can remember the last voltage that passed through it.[25] When Di Ventra delivered external voltage in a pattern, similar to how Nakagaki had exposed the slime mold to alternating hot and cold temperatures, he found that the memristor circuit could predict voltage fluctuations, just like *P. polycephalum* predicts temperature fluctuations.[26]

That result could be tremendously important in bridging the gap between human and artificial memory. Most computers, after all, are built to perform one operation at a time—and the faster our processors have gotten, the faster they've been able to stack up these operations in their memory. But human intelligence works differently, with synapses getting reinforced over time like the channels in the slime mold's cytoplasm.[27]

That difference in architecture is tremendously important when it comes to energy consumption. When it sought to simulate a human brain—with 10 billion neurons and 100 trillion synapses between them—IBM employed its famed Sequoia supercomputer, which takes up 3,000 square feet of lab space at Lawrence Livermore National Laboratory. Lauded as "the pinnacle of energy efficiency"[28] as supercomputers go, Sequoia nonetheless needed 7.9 megawatts to run, or enough energy to power thousands of homes at once. The human brain, on the other hand, needs just 20 watts—enough to power a low-energy light bulb.

Oh, and that simulated brain? It might have achieved the computing *power* of a human brain, but not the speed. It ran 1,542 times *slower* than the organ that sits in your skull.[29] If we can get human-made data storage and processing units to work more like natural data-storing and processing units, we could make a big dent in our spiraling energy needs.

That might be especially important in a world of cryptocurrencies. The electricity it takes to "mine" and use the world's first and most popular decentralized digital currency, Bitcoin, produces a lot of emissions—perhaps enough to provoke a 2 degrees Celsius rise in global temperatures by 2033, according to researchers at the University of Hawaii at Manoa.[30]

"We need to rethink the way we compute," bio-inspired computing expert Julie Grollier told the World Economic Forum in 2016. "Our computers are excellent at precise arithmetic; our brains excel at cognitive tasks." When it comes to recognizing patterns without being told ahead of time what sort of patterns to expect, she noted, organic intelligence is still far superior.[31]

In a paper published the following year, Grollier and her team, based at the National Center for Scientific Research in France, showed that circuits employing memristors acted more like a human brain, processing and storing information in parallel instead of running through one computation at a time. The result is the ability to recognize structure in unlabeled data and respond accordingly—just like the slime mold did.[32]

Ultimately, what we're learning from intelligent single-celled organisms may help us mimic human-level smarts with artificial intelligence. But if we were truly as wise as we like to believe, we wouldn't just be looking to mimic the way *individual* organisms think. There's power in numbers, after all.

HOW ANTS ARE TEACHING US TO NAVIGATE A COMPLEX WORLD

Ant brains are really big. At 15 percent of body mass in some species, like the "rover ants" of genus *Brachymyrmex*, they're the biggest in the world in terms of relative size—which is often used as a rough, albeit imperfect, proxy for intelligence.

Ant brains are also really small. *Brachymyrmex* brains have been measured at .005 milligrams—about one-tenth the weight of a single grain of salt.[33]

However you measure, this much is clear: Ant brains are incredible things. Because even though some ant species' brains are many millions of times smaller than ours, they can think in ways that make us look downright stupid by comparison.

When ants work together—and, as we know, one of the things that makes ants so tough as a species is that they are almost always working together—they can process lots of information very quickly, building order from chaos in ways that maximize the fast, efficient, and life-sustaining gathering of food.

An individual scout ant's actions might seem random, cutting this way and that, backtracking, spinning around, and zigzagging about in the long and often unsuccessful quest to find a food source. But part of that perceived chaos is simply that we have a hard time appreciating the world from an ant's point of view. Even though ants are great climbers and their six legs and strong pinchers allow them to scale large obstacles, it's often easier to find a way around something than to go over it—and so an individual ant's job isn't just to find food, but to evaluate all of the different permutations of pathways leading to and from that food, resulting in the least energy output from other ants once food is found.

Once a single ant locates something yummy, that ant takes a little piece and heads back to the nest. Along the way, it secretes pheromones, marking the way back to the food. But because the pheromones of one single ant don't last very long, the first few ants to follow the trail still wander around quite a bit as they try to pick up the scent. Over time this wandering results in better and more direct shortcuts the first ant didn't identify. And as more ants join the parade, the pheromone trail becomes like a well-lit and very direct freeway from nest to food and back.

All of this works a lot like an internet search engine, where web crawlers travel about the internet, examining all of the nooks and crannies of the ever-changing World Wide Web, in search of new morsels of information. When Google's crawlers find something new, that data is brought back to be indexed, and the crawlers are sent out again and again, with a greater focus toward evaluating all the possible paths to that information, and ensuring the information is still there and can be accessed efficiently.[34]

Google indexes tens of trillions of web pages, hundreds of billions of times each month.[35] But to mathematician Jurgen Kurths, the ways in which ants turn seemingly random search patterns into a linear path makes Google's illustrious algorithms look like elementary school arithmetic.

"That transition between chaos and order is an important mechanism and I'd go so far as to say that the learning strategy involved in that is more accurate and complex than a Google search," Kurths said in 2014. "These insects are, without doubt, more efficient than Google in processing information about their surroundings."[36]

Kurths isn't just any mathematician. The professor of nonlinear dynamics at Berlin's Humboldt University is one of the most influential complex-systems scientists in the world. His work demonstrating the "chaos-order transition" in ant foraging behavior, published in 2014, has already influenced scientists developing computing systems designed to simulate biological neuron networks,[37] working to solve complex mathematical optimization problems,[38] and reducing uncertainty in computational modeling.[39]

Ants aren't just good at working together to build floating rafts and foraging freeways, though. They're also quite adept at building bridges—literal ones, using their own bodies. And that's not just selfless; it's smart.

In fact, it's *really* smart, because they do it without any apparent leadership. Imagine, for a moment, what the Brooklyn Bridge would look like if Washington and Emily Roebling, instead of directing the work of the bridge's construction crews in the late 1800s, simply told each worker, "Do what you think is best." Yet that's exactly what the army ants of the species *Eciton hamatum* do when confronted with the need to cross gaps in the forest floors of Central and South America.[40]

And if the gap expands, the bridge does, too, again without any apparent central direction—up until the point at which the ants make a collective decision that they've devoted too many workers to the bridge and need a change in plans. Then, as one, they begin deconstructing the bridge and going back to finding a new path to food.

"These ants are performing a collective computation. At the level of the entire colony, they're saying they can afford this many ants locked up in this bridge, but no more than that," biologist Matthew Lutz said in 2015. "There's no single ant overseeing the decision, they're making that calculation as a colony."[41]

While the idea that ants are collectively intelligent is gaining greater acceptance, there are a lot of folks—even those who study insects—who don't think much of an individual ant's brain power. Even Kurths believes that, while ants collectively act in ways that mimic what we think of as intelligence, "the single ant is certainly not smart."[42]

But I have to rise in defense of the little guy. Because much as you can't measure human and dolphin intelligence in the same way—even though, in the grand scheme of evolutionary things, we're really quite similar—it would be crazy to measure ants with a human IQ test. Yes, that has a lot to do with the fact that ants have off-the-charts collective intelligence. But new research is demonstrating that, even at an individual level, ants are spectacularly intelligent.

Just consider what they have to do to navigate the enormous world around them before those slick highways and bridges are built. Ants don't see the world like we do; most ant species' eyes are good for detecting movement, but not for making out shapes and measuring distances.[43] Yet in almost every measurable way, they're far better than we are at figuring out where they are in the world at any given moment.

When humans navigate from one place to another, we almost always rely on visual clues, with a bit of auditory stimulus thrown in for good measure and, if we're searching for a place that makes cookies, hamburgers, or pizza, maybe a smidgen of olfactory information, too.

Individual ants collect and process a lot more cues than we do. Like us, they use the shape and size and movements of structures around them to take stock of where they are and figure out where they are going, but they also use the position of the sun, polarized light patterns, the wind, minute changes in odors, the feel of the ground under their feet, and even the number of steps they've taken since leaving the nest.[44]

Want to navigate like an ant? Start on one side of a city park and start walking, keeping track of the number of steps you take along the way. Every time you run into an obstacle or a change of terrain—like a picnic bench or concrete pathway that crosses the park—remember that feature and where it was in relationship to the last feature. Keep the position of the sun in mind, too, and when you step into the shadow of a tree, remember how long the shadow lasted before you were back in the sun again. Which direction is the wind coming from? Make sure you notice that. What odors are being carried on the breeze and where are you when you smell them? Remember that, too.

Oh, and now imagine that at any given time a hungry creature hundreds of times your size might appear from out of nowhere and devour you. But don't let *that* get in the way of memorizing all of the things you need to remember!

There aren't many humans who could simultaneously process all of this information. The human brain directs navigation by taking bits and pieces of sensory cues, mostly visual ones, and using those pieces to create a cognitive map—a mental representation of our environment, often in relation to where we are and where we want to go.[45] To create this map, though, we have to discard a lot of information, and since most of us rely so heavily on visual cues, the information we discard tends to be all of the other stuff. That's why we get lost so easily in the dark, and why we bump into things in our own homes when the lights are off.

Ants don't create a cognitive map. Instead, they maintain multiple memory modules that they can access, either singularly or in combination with other modules, when a situation arises.[46] And since they're not reliant on any one representation of the world, they're a lot less likely to get lost when the world around them changes; if it gets dark, a major obstacle shows up or is removed, or the wind starts blowing a different direction, they simply access a different module.

All of this is quite handy for humans trying to solve one of the most complicated engineering problems in the world: self-driving cars. For the most part, driverless vehicles work great on well-mapped and well-marked highways. But they work a lot less great in dynamic cities and construction zones. Increasingly it is becoming clear that the best way to handle the mostly predictable but occasionally anarchistic world of automotive travel is to rely both on collective and individual intelligence.

When it comes to reducing traffic, for example, collective intelligence is key. By studying the ways in which ants keep from jamming up, even as their numbers increase,[47] physicist Apoorva Nagar has observed that three simple rules keep ant traffic moving along without gridlock. First, since ants don't have egos, nobody is trying to get ahead of anyone else, and nobody gets pissed off when they're overtaken. Second, they don't stop when they bump into one another, so minor fender benders don't interrupt the flow of traffic. And third, the more congestion there is, the straighter and steadier they travel.

It's easy to see how rules one and three could apply to self-driving traffic. Rule two seems a little more complicated to apply to humans and cars—until you remember that it doesn't take an *actual* collision to get human drivers to swerve or slam on the brakes, which tends to have a cascading effect that can slow down every car behind them for many minutes and several miles. Close calls don't need to slow traffic if whoever or whatever is operating the car doesn't swerve or brake any more than absolutely necessary; you could call this the "no blood, no foul" rule of self-driving vehicles. The sort of precision and discipline it would take for these rules to work is only possible if *every* vehicle follows the rules, as ants and algorithms do.

However, as the pioneers of self-driving technology learned throughout the 2010s, such simple rules become worthless when something like construction comes into play. When road cones replace traffic signs and hard-hatted sign holders replace signal lights, an individual vehicle, like an individual ant, needs to follow a far more complex set of rules. And since we live in a dynamic world, one intelligence module simply isn't enough to determine what those rules should be or how they should be followed.[48] GPS is a good start for navigation until satellite signals are lost. Cameras are worthless when they're covered by snow. Laser distance measurements are accurate when it comes to stationary objects but not as good for moving ones. Radar can be more or less valuable than sonar depending on the conditions of the environment. And maps are great until people in reflective vests show up in the middle of the road. To succeed in these sorts of environments, self-driving cars can't rely on a unified map of the world, but rather must independently take information from multiple available modules, use it to solve the problem at hand, and communicate the results to the community—just like ants do.

It takes a good dose of humility to accept the fact that a tiny arthropod is, both singularly and in concert with its colony, so much smarter than we are. But once we accept the fact that we can learn a lot from ants and other really smart animals, we can start taking advantage of what they have to teach us. It also makes it easier to take another really big cognitive leap: accepting the intelligence of life-forms that, up until very recently, weren't thought to have any intelligence at all.

WHAT WE'RE LEARNING FROM SMARTY PLANTS

"If you sing to the plants when you water them," my grandmother told me when I was five, "they'll grow better."

I had no reason to doubt her. She'd never lied to me before, except for that whole Santa Claus thing, and that was forgivable because she had brought me a *Star Wars* rebel transport ship when I was four, when all my other cousins got much smaller spaceships or single figurines.

So I sang to her plants when I visited her home. And I sang to the ones we had at my family's home. And this went on until I was in the second grade when, having had the popsicle stick with my name on it drawn from a construction-paper-covered coffee can, I was selected as that week's "class gardener."

While watering the classroom ficus that afternoon, I sang, to the tune of Allan Sherman's "Hello Muddah, Hello Fadduh":[49] "It's gotten hotter. So here's your water. Want to thank me? Please don't bother."

I remember this day because it was emotionally damaging. My classmates laughed at me. My teacher seemed concerned. And when I explained, matter-of-factly, that singing makes plants grow better, my classmates laughed ever harder and my teacher looked even *more* concerned.

My grandmother hadn't been lying to me. Not really. She was one of millions of Americans in the mid-1970s who, thanks in no small part to a book called *The Secret Life of Plants* and the ubiquity of the New Age movement, came to believe that plants might be perceptive and responsive to human words, songs, and even thoughts. The book had its roots, so to speak, late one night in 1966 when a polygraph examiner named Cleve Backster decided, apparently on a sleepless whim, to hook his lie detector up to a dracaena, a common house plant often sold as a "dragon tree." Wanting to stress the plant, Backster dipped one of its leaves into a mug of hot coffee and, when nothing happened, he conceived of a worse threat—he'd burn the leaves.

"The instant he got the picture of flame in his mind, and before he could move for a match, there was a dramatic change in the tracing pattern on the graph," authors Peter Tompkins and Christopher Bird wrote. "Could the plant have been reading his mind?"[50]

The answer to that question is an emphatic no. Scientists who have tried to replicate that and other examples of Backster's purported findings—which included the supposed discovery of inter-eukaryotic empathy as Backster dropped live shrimp into boiling water in front of his photosynthetic test subjects—have failed. But that didn't keep *The Secret Life* off bookstore shelves, nor did it prevent a documentary of the same name from being produced, with a stunningly beautiful soundtrack by Stevie Wonder, no less.[51] And for a long time to come people like my grandmother believed that being nice to plants—by serenading them, for instance—would help them grow happier and healthier, much to the annoyance of the scientific community.

It was several decades before most serious scientists were once again willing to publicly discuss the possibility that plants might collect, process, and share information—this time not through some magical form of telepathy but through complex phytobiology.

Among the pioneers and most passionate advocates in the field of plant signaling and behavior is Elizabeth Van Volkenburgh, a University of Washington plant biologist who was one of six authors on a 2006 paper in the journal *Trends in Plant Science* that announced the new field of "plant neurobiology."

The pushback on that paper was immediate and intense, with twenty-six scientists signing onto a letter in the same journal, a few months later, accusing Van Volkenburgh and other proponents of plant neurobiology of making "erroneous arguments" in launching a field "founded on superficial analogies and questionable extrapolations" that lacked "an intellectually rigorous foundation."[52]

"What was surprising to me is the resistance," Van Volkenburgh told me. "And then there was the resistance to the resistance. It seemed much more cultural than scientific."

It wasn't just animal biologists who had a problem with the idea of plant neurobiology, Van Volkenburgh said. "A lot of scientists, especially those who teach in plant biology programs, didn't want to reduce the magic and majesty of plants into something so lowly as humans," she laughed. "That's an argument that I could really understand, although I hadn't thought of it that way myself. I never thought we were making plants as primitive as humans; I thought we were complementing humans by saying that animals had something in common with plants."

More than a decade later, the debate was still lively, even emotional.[53] The first time we spoke, Van Volkenburgh had just finished meeting with a young graduate student from the University of California at Davis who was hoping to study the ways plants communicate.

"That sounds like a brave young scientist," I said.

"She absolutely is," Van Volkenburgh replied, "and she'll have to be."

But Van Volkenburgh was also optimistic that an evolution was underway in the way we understand plants. Because whether you call it neurobiology or intelligence or something else, it's getting harder and harder to pass plants off as vacuous beings.

Take Monica Gagliano's research subject of choice, the delicate *Mimosa pudica*, also known as the touch-me-not, which defensively folds its leaves when touched. A few years ago, Gagliano got to wondering whether she could teach the touch-me-not to stop curling its leaves. So she loaded a bunch of the plants, one at a time, into a special plant-dropping contraption and dropped them, again and again, from a height of 6 inches. Somehow, over time, the plants seemed to figure out that being dropped in this specific way wasn't going to cause any actual harm, and they stopped furling their leaves.[54]

If Gagliano disturbed the plants in other ways—like shaking them—they still engaged in their trademark defensive behavior. But even four weeks later, the plants still appeared to remember that the six-inch drop was no cause for alarm.[55]

Is that memory? Before answering that question, Van Volkenburgh likes to challenge her students to consider another question: How does the human brain remember things?

"Sometimes my students will have very recently taken a course in human biology, and they're often surprised to realize that they didn't learn the answer to that question," Van Volkenburgh said. "That's because nobody knows how brains remember things. And that's often an eye-opening moment."[56]

Clearly, plants don't have brains. But just because they don't have the organ we use to collect, store, and process information doesn't mean they can't do it. Fish don't have lungs, after all, but they still breathe.

And increasingly, it's becoming clear: Plants can *learn*.

In an experiment that utilized one of the most famous plants in the history of science—*Pisum sativum*, the same garden pea Gregor Mendel used to establish rules of heredity—and built upon one of the most famous animal studies—Ivan Pavlov's experiments on canine conditioning—Gagliano demonstrated that plants can learn by association just like animals can.

As anyone who has ever seen a windswept tree on the coast can attest, plants will generally grow away from a stiff breeze. But when Gagliano and her research team coupled a fan with a light source, plants grew toward the wind. That by itself isn't all that surprising—for a pea plant, the need for light is paramount. But when the team took the light away and moved the position of the fan, the plants that had been conditioned to associate wind with light still grew toward the breeze. Associative learning—a fundamental building block of just about any definition of intelligence—appears to be a universally adaptive mechanism shared by animals and plants alike.[57]

But where is this sort of decision making coming from? One big clue arrived in 2017, when scientists identified a group of a few dozen cells in a common roadside weed called *Arabidopsis thaliana*, or thale cress, that appear to control the process by which seeds decide to sprout.

That's right—decide. The cells are separated into two subgroups. The first group produces a lot of the hormone gibberellin, which promotes germination. The second group produces plenty of abscisic acid, which promotes dormancy. Both hormones are impacted in antagonistic ways by temperature fluctuations, and these signaling molecules are sent back and forth within the cellular group until conditions are such that the germination promoters prevail.[58]

What else are plants deciding? How much water to consume? What specific nutrients to pull from the soil? How much strain and stress to endure in exchange for sunlight? We're only at the cusp of finding out.

So what's the world's smartest plant? There are nearly 400,000 identified plant species, which collectively make up more than 80 percent of the world's biomass.[59] Researchers have only studied a handful of them. It's going to be a long time before we understand the myriad ways in which plants are intelligent, let alone identify the Einsteins among them.

In the meantime, our attitudes about plants are almost certainly going to be affected in complicated ways. The vast majority of life—some 450 gigatons of carbon worth—is made up of plants, after all. "Where does all this lead us?" writer Prudence Gibson asked in 2016. "Well, into troubled waters, so grab your boat and paddle. We are in for a rough philosophical ride."[60]

Van Volkenburgh certainly hopes so. In evolutionary terms, we're all eukaryotes—plants, animals, and fungi didn't diverge until about 1.5 billion years ago. Up to that point, we were on the same genetic road, having shared a lineage for about 2 billion years. And we share more than that: Our cells are markedly similar, with nuclei, cytoskeletons, cytosol, peroxisomes, Golgi complexes, cell membranes, endoplasmic reticulum, lysosomes, and mitochondria. We have long, shared sequences of DNA, which is constructed in the same double helix shape and comprises the same four nucleotides. "We have far more similarities than differences," Van Volkenburgh said. "And maybe if we realized this, we might treat our planet differently."

Maybe plants don't think the way humans do. But neither do elephants, dolphins, octopuses, ants, or amoebas.

We do ourselves absolutely no harm by acknowledging that all of these creatures are intelligent—in ways that we can barely comprehend, no less—and that they all have plenty to teach us about living on Earth.

Conclusion

THE NEXT SUPERLATIVE DISCOVERY IS YOURS

About the same time I was getting to know Zuri, the elephant calf, my wife brought home a big red children's book, Steve Jenkins's *Actual Size.*

The book ostensibly was for our child, a little girl we call Spike. It somehow wound up on our coffee table, though, and I found myself regularly leafing through it, gazing at Jenkins's beautiful, life-sized collages of a variety of remarkable animals. There was the world's largest lepidopteran, the Atlas moth, whose wingspan is nearly a foot across. There was the largest bird in the modern world, the ostrich, which can stand more than 9 feet tall. There was the world's largest reptile, the saltwater crocodile, which can grow to a length of 25 feet.

These creatures, and the many other organisms of superlative distinction in the world, are more than just spectacles to be gaped at. They are ambassadors for conservation. They are clues leading us to a deeper understanding of our world. They are the source of excellent, actionable knowledge that can be used to improve and even save lives. They are connectors to the interconnectedness of all living things.

And a great many of them have yet to be discovered.

Everyone can do science. *Everyone.* Many of us learned this while preparing grade-school science fair projects, for which we embarked upon noble quests to determine which ice cream melts the slowest, to understand which sort of fish cats find tastiest, or to figure out which fruits and vegetables generate the most electricity. At some point most of us stopped asking questions like that.

That's what happened to me. For a long time it felt like scientists had already answered all the simple questions of the world. And the mysteries that remained were so complicated that I couldn't even formulate a question about them. Just like that, my awe was gone.

Yet I remained enamored by extremes, as a lot of people do. And this, I think, presents us with an opportunity to re-engage in the world with the awe and excitement of children—for the next superlative discovery could well belong to one of us.

What would you like to discover? What long-ignored outlier? What unsought extremophile? What marvel of biology?

Perhaps you would like to discover the loudest creature at some level of taxonomy. The loudest fish, perhaps? The current record holder, a croaking fish called the Gulf corvina, *Cynoscion othonopterus*, has been recorded at greater than 175 decibels in Mexico's Colorado River Delta.[1] But more than 1,000 fish are known to make noise. Most of them haven't been studied. And making an initial survey is as simple as dipping a cheap hydrophone into the water the next time you go fishing.

Maybe you're interested in the toughest of creatures? New tardigrade species are being found all the time, and chances are good that if you live in a place where there is moss, you have tardigrades living near you. Finding them is a rather simple matter of collecting a small clump of the moss, putting it in a Petri dish, dropping some water onto it, waiting a day, and then looking at the water under a microscope.

How about the fastest animal on our planet? Recall that the current record holder for relative speed, a tiny mite called *P. macropalpis*, was hiding in plain sight—it had been identified more than a century ago but was never really studied until an undergraduate student took notice, collected some video footage, and did a bit of math. And remember that no one knew how fast the record holder for absolute speed, the peregrine falcon, could go, until an amateur scientist and recreational skydiver decided to find out.

What about the tallest of some taxon? It wasn't until 2006 that the world's tallest known singular tree, a sequoia in California, was discovered—and it's certainly possible that another, taller redwood might have thus far been missed, as the 379-foot Hyperion was for so long. In the summer of 2016, after all, scientists announced the discovery of the world's tallest tropical tree, a nearly 300-foot yellow meranti in Borneo, only to be outdone by the discovery of not one but fifty taller meranti elsewhere on that same island. Very precise measurements can be tricky, of course, but a little right-angle trigonometry could send you on the path to superlative discovery.

Or perhaps you might like to make the *biggest* discovery of all? I certainly would.

The genetic study that established Pando to be a singular aspen clone wasn't the last word on "largest." For the moment, at least, Pando is still the largest known plant. But not far to the northwest of the great aspen is another contender for the biggest of all living things, one that has proven even harder to measure—an organism that, like the mighty aspen clone, is largely subterranean.

They call it the "humongous fungus," a genetically singular specimen of *Armillaria ostoyae* that, at nearly 2,500 acres, covers some twenty-five times the ground Pando does. Creeping underground and penetrating the roots and bark of the trees it feasts upon, it waxes and wanes with the season, blooming with millions of mushrooms for a few weeks each fall. It's unclear whether it is fully interconnected, but some researchers have speculated that, if it is, and if you gathered it all together at certain times in its cycle, it could be heavier than Pando, too. There's a lot of research left to be done.

For a long time, I figured the "great unknown" was what had made Burton Barnes, the first person to map out the aspen clone that later became known as Pando, so reluctant to take credit for his massive discovery.

When I asked him to explain in 2013, though, he demurred.

"Many ideas and comments were written about the large clone and other candidates, but I was not inclined to make a contribution at that time," he told me of the period that followed the article in *Discover* that introduced Pando to the world. "That's still my view today."

For years I missed the clue he'd offered me. And then, one summer afternoon, reclined in a hammock in Pando's southeast corner, where the world's largest known organism meets a much smaller cousin, it struck me.

"The large clone," he'd said, "*and other candidates.*"

We always talk about Pando—or should, at least—using the word "known." There are always unknowns, after all, and thank heavens for that, for it is the mysteries of the natural world that make biology such fun. Yet the idea that Pando could be superseded by another, bigger, clone was always a bit nebulous to me—until that moment. Looking up at where Pando's branches met the other clone, I could make out a distinct contrast between one and the other. Blurring my eyes, a bit, I could see that the leaves were subtly different shades. Focusing on the stems, I could see that Pando's branches twisted and bent in a way that was different than its cousin's.

I shot up in my hammock so quickly that I nearly came tumbling out. I packed my things and began to move up the mountain, looking down every few steps to see if a bar had appeared on the screen of my phone. I hiked and looked and hiked and looked.

Finally, gloriously, one bar. And then, miracle of miracles, two bars. That was good enough.

It wasn't easy to find Barnes' original article on the clone, but with a little help from my university's librarians—who are superheroes, by the way—it popped up on my little screen.

Here's what Barnes had written: "Large aspen clones, from 10 to 200 acres, are not uncommon in the central and southern Rocky Mountains."

Two hundred acres.

Two hundred acres?

Two hundred acres!

Barnes didn't make much of his 100-acre Fish Lake clone because he didn't think it was actually superlative. The guy who discovered the organism that came to be known as Pando thought there were other, bigger aspen colonies out there.

Where? Barnes took that secret to his grave.

But wildland resources researcher Paul Rogers believes it's not only possible but entirely likely that there is a larger clone out there. "Aspen are so widespread," he told me as we stood in a thick section of stems I've come to think of as Pando's heart. "This one just happens to have a road running right through it. If the road had gone a different way through this basin, we might have never known it was here."

The hunt for other massive aspen could be elaborate, expensive, and hard. It could involve drones, infrared cameras, large teams of researchers, and DNA collection protocols like those Karen Mock used to confirm Pando's size.

Or, it could be much simpler. After all, Barnes identified the Fish Lake clone with little help and absolutely nothing high tech.

And you could do that, too.

A remarkable graduate student named Kristina Flesher demonstrated how simple the process can be in her master's thesis at Michigan Technological University, in which she sought to understand just how good "phenotypic delineation" was compared to DNA-based mapping. In doing so, she showed that Barnes' seemingly uncanny ability to identify individual aspen wasn't magic. It was something that anyone could get good at, if they were determined to do so.[2]

If you have a grove of *P. tremuloides* near you (and if you live in the northern or western United States, or anywhere in Canada or Alaska, you probably do), all it takes is a walk in the woods. And even if you don't live in those areas, there are other species of aspen—*P. grandidentata* in the North American east; *P. adenopoda*, *P. davidiana*, and *P. sieboldii* in central and eastern Asia; and *P. tremula* in northern Europe and western Asia—that also grow as clones, and that have not been mapped nearly as extensively. Not yet, at least.

For reasons that are not only scientific but also life-affirming, one of the best times to go see aspen is during the fall, when the leaves are changing and the entire canopy is aglow. Leaves from individual clones, even those that might be related, change colors at different times and in different ways, and in these moments the boundaries separating one clone from another can be quite stark, a sea of pale yellow set against an ocean of vivid green, a field of bright orange next to a range of glowing gold.

At any time of the year in which aspen have leaves, though, the color can be a subtle clue that different stems belong to the same clone. Before one recent trip to Pando, I stopped by the hardware store down the street from my home and picked up a stack of paint color cards. The difference between "Windsurfer" and "Mint Truffle" may be subtle, but it can also be a very significant clue that you're looking at leaves from two different clones.

The proportion of colors on a leaf can also offer important information. Are the leaves all green? Green with a little yellow? Half green and half yellow? Yellow with a bit of brown?

As you might suspect, the shape and size of leaves can tell us a lot about belonging, too. As Barnes first observed in his doctoral dissertation at the University of Michigan in 1959, the length-to-width ratios of leaves from the same clone are remarkably consistent. Barnes also observed, and Flesher elaborated upon, the ways in which leaf serration could be used to separate leaves from different clones. In *P. tremuloides*, Barnes noted, some leaves had as few as twenty teeth per side, while others had nearly twice as many. For consistency, Flesher suggested counting only the serration tips on the left side of each leaf, starting at the stem and moving clockwise to the tip.

Is it springtime? Take a look at the leaf buds. Are they closed and brown? Have they begun to break out with pointy shoots of green? Is the leaf itself beginning to unfurl? Is it fully open? Different clones emerge at different times.

Bark is also telling. Aspen usually look starkly white at first glance, but if you hold a grayscale color card—or one with shades of pale green, brown, orange, and yellow—up to the stem, you'll notice a huge variation in bark color, especially if you first wipe away the chalky white powder common on many aspen (that's the tree's sloughed-off dead cells, by the way).

None of these variables by themselves will tell you whether you're looking at stems from one or multiple aspen, but if you put them together, it becomes much easier to see the contrasts between one clone and another.

I firmly believe that there is something greater than Pando out there. Finding it is just a matter of time, patience, and luck.

It is late in the winter here in the Wasatch Mountains of northern Utah, and from the window where I am standing, writing these words, I can see the spindly skeletons of aspen stems, stark against the snow-covered mountainside. In just a few weeks, the leaves will begin to emerge, and I will walk in the forest and watch for the contrasts.

I will search for something that is greatest, and no matter what comes of that search, I will find something great.

ACKNOWLEDGMENTS

This book would not have happened if Paul Rogers had not agreed, many years ago, to introduce me to a one-tree forest called Pando, and I wouldn't have been inspired to write about it, and other superlative life-forms, if Paul hadn't modeled for me a particular kind of reverent, joyful awe that I now try to share with others.

Paul is one of hundreds of scientists whose studies are referenced in this book. All of those studies cite other studies. And on it goes. Science is a collective enterprise, and it's turtles all the way down. Thus, people who write about science are indebted to more people than we could ever calculate.

How, then, can I properly thank all of the researchers who made this book possible?

One thought: When people find out I'm a US military veteran, they often say, "Thank you for your service." Having done nothing to warrant anyone's thanks, I wasn't a huge fan of that phrase. Then my friend, Jared Jones, a military helicopter pilot who has completed several tours of duty in Afghanistan, told me he doesn't figure people are really thanking *us* for our individual service. "I think most people realize that not everyone in the military is an actual hero," he said, "but since there's no way to know who we're actually indebted to, it's easier just to thank everyone." And that's why, these days, when I meet a scientist from *any* field, I say, "Thank you for being a scientist."

I am also quite deeply indebted to the army of science writers whose work I often relied upon to ensure that my reading of a particular piece of research was sound. It often wasn't, at first, and it is people like Kristin Hugo, Janet Fang, Alex Zielinski, Tracy Staedter, Rebecca Boyle, Carrie Arnold, Hayley Bennett, Henry Nicholls, Mark Strauss, Sascha Steinhoff, Sid Perkins, Brendan Buhler, Erica Goode, Kate Tobin, Brian Switek, Janet Raloff, Laura Helmuth, and the brilliant and prolific master of the field, Ed Yong—among so many others—who have helped me "get it."

On the subject of getting it: Sometimes, even when scientists have patiently taken me step by step through their process and findings, I don't see things the way they do. Any confusion, or outright mistakes, regarding the work of the researchers who I have written about in this book are my responsibility alone.

This book was written in coffee shops around the world, but in one more than any other. I'm not sure how much I've spent on coffee at Alchemy, in Salt Lake City, but I know this: It's not nearly enough to repay the hundreds and hundreds of hours I have spent in an establishment that has become one of my favorite places in the world to write. The cast of characters—customers and staff members—who rotate in and out of the funky little café on 1700 South bring me endless joy.

I would not have been able to write this book were it not for the support and flexibility granted me as a faculty member at Utah State University. I am forever in debt to Professor Ted Pease for inviting me to Logan, and for modeling for me the sort of professor I would like to be; and to former dean John Allen, who believed that people who do, teach. I am also most grateful to my colleagues in the Department of Journalism and Communication, who inspire me every day—and in particular to Candi Carter Olson and Debra Jenson, who listen to me, support me, and uplift me in ways I fear I cannot ever repay. And no discussion of the support I get from my university would be complete without mention of my students and former students, who push me to be worthy of the way I have pushed them. The stories of our world are in their hands, and everything is fine.

I likely would not have ventured anywhere near science if Terry Orme, then the managing editor of the *Salt Lake Tribune*, hadn't relented to my pestering and permitted me to cover Utah's Hogle Zoo. I am also grateful to the series of editors I have had since starting my writing career some twenty years ago, who encouraged me to pursue subjects that were beyond my actual capabilities: A. K. Dugan, Steve Bagwell, Steve Fox, Brent Israelsen, Greg Burton, Tom Harvey, Joe Baird, Rachel Piper, and especially Sheila McCann. It is a testament to all of their work, over the years, that my editor at BenBella, Leah Wilson, is always commenting about how clean my work is when it arrives in her inbox.

From the moment I signed with BenBella, I knew I wanted to work with Leah. I am so proud and so fortunate that she opted to partner with me on this book. Her thorough, thoughtful work—right down to the reference notes—elevated this project beyond anything I could have hoped for, and her cheeky margin comments made me smile every day.

Leah is exemplary of the team at BenBella, whose kindness, expertise, and professionalism makes me want to be nicer, smarter, and better at what I do. I have greatly enjoyed working with Adrienne Lang, Lindsay Marshall, Alicia Kania, Susan Welte, Monica Lowry, and the entire crew from this amazing organization. I am deeply grateful to copy editor James Fraleigh, whose attentiveness to detail—and encyclopedic knowledge of everything from the Hawai'ian 'okina to Star

Wars spacecraft—not only prevented some embarrassing mistakes, but also taught me a lot about the world. I also appreciate the work of Sarah Avinger, who led the team that produced a cover that made me cry when I first saw it. And I greatly admire Glenn Yeffeth, the man whose vision, decency, and humor made such an organization possible.

I would not have found BenBella were it not for Trena Keating, of Union Literary, who has never made me doubt that I am her favorite and most important client (even though I know who her other clients are, and thus I know I cannot possibly be her favorite or most important client).

And I would not have found Trena if not for John Day, who trusted me to work on his first book, *Longevity Plan*, and with whom I am looking forward to a long lifetime of collaboration and friendship. I am grateful, also, to the other authors with whom I have collaborated. Sharon Moalem believed I could write about epigenetics even after I explained that I knew nothing about epigenetics. And David Sinclair, whose work will change the world, is a good bloke and a good mate; I am proud to say I am his co-writer, but so much prouder to say I am his friend.

On the subject of friendship: This book never would have happened were it not for the support of Scott Sommerdorf and Matt Canham, who taught me that it was OK to think about things other than newspaper journalism; that mostly meant poker, at the time, and there's nothing wrong with that. I likely would not have made the jump from a life in the military to a life as a writer if it were not for Roger Weaver, and I strive to be as good a mentor to my students as he was to me. Even then, it was the support of Katie Pesznecker, Scott Johnson, Troy Foster, Jennifer Joan Nelson, Andrew Hinkelman, DeAnn Welker, Jake Ten Pas, Carole Chase, and Joel Fowlks that made the transition possible. My next major life change, from full-time journalist to full-time academic, would almost certainly have stunted my growth as a writer if not for the people of CReEL, who meet each year in A. B. Guthrie's cabin in Choteau, Montana, to celebrate and elevate one another's work. To my dear friends and fellow founders, Alex Sakariassen and Bill Oram, and my fellow attendees over the years, Gwen Florio, Camilla Mortensen, Aaron Falk, David Montero, Casey Parks, Emily Smith, and Jaime Rogers, I do not have enough words to describe how much you all inspire me. Two other members of the CReEL crew deserve especial gratitude, not just for being inspiring writers and brilliant editors, but for the deep and generous friendship they gave to me during the year I wrote this book, which—for reasons that had little to do with writing this book—turned out to be the hardest of my life. Sarah Gailey and Stephen Dark: I love you.

And, as long as I'm writing those three words, let me express them also to my mother, Linda, and my father, Rick; my sister, Kelly; and my brother, Mikey. I love you.

And Heidi Joy, to whom this book is dedicated: You have endured a lot to be with me, and I will never be worthy of your humor, grace, beauty, wit, and intelligence, but I will continue to try. I am so proud to be your partner.

And finally, to Mia Dora, who still lets me call her Spike: You are the smartest, bravest, most beautiful, toughest, and kindest person I have ever known. You are Superlative.

NOTES

INTRODUCTION: NATURE'S BEST AMBASSADORS

1. Zuri didn't yet have a name when we first met. "For now," I wrote in the *Salt Lake Tribune* on September 8, 2009, "you might just call her Exuberance."
2. Guinness has long contended that it is Taman Negara National Park in Malaysia, which is estimated to be 130 million years old; others have suggested Daintree Rainforest in Australia, which might be 135 million years old and perhaps older than that. The debate offers excellent opportunities to explore the question of "oldest" from many points of view and, in doing so, strengthen our scientific understanding of both of these amazing places. That's one of the things I'm hoping to do in this book, too. Resolving subjective debates is pointless. But we can use these debates to expand our understanding.
3. The bullet ant, tarantula hawk, and warrior wasp all register a 4 on the 4-point pain scale developed by entomologist Justin Schmidt, who wrote about his work comparing the impacts of stinging insects on humans—mainly using himself as the test case—in a wonderfully whimsical book called *The Sting of the Wild.*
4. Probably the diplodocus, which might also have had the "fastest" tail, as well. One computer model suggests that the great sauropod's tail could reach supersonic velocities, making a noise like the cracking of a bullwhip.
5. That's what *Guardian* science editor Robin McKie calls our ancestors. She also noted that what we now call "trophy hunting" might be deeply ingrained in our evolutionary history. In "Humans hunted for meat 2 million years ago," McKie wrote about a 2012 discovery, by anthropologist Henry Bunn of the University of Wisconsin, that pushed back the date of systematic human hunting by hundreds of thousands of years.
6. I came to know about *Conraua goliath* from Steve Jenkins's excellent children's book,

Actual Size. Jenkins is brilliant. His other titles include *Biggest, Strongest, Fastest* and *Apex Predators*, so he is clearly as enamored by superlative creatures as I am.

7. I often mock esoteric paper titles, but seemingly esoteric research can be extremely important. This paper, by Jie Xie, Michael Towsey, Jinglan Zhang, and Paul Roe, came as a result of the recognition that frogs, while excellent indicators of the state of the environment, can be hard to count. Automated frog call classification is an important part of monitoring frog populations.

8. This study showed that metal pollutants have different impacts on different frog tissues, and demonstrated that the liver and skin are better for assessing metal-induced oxidative stress than muscle is. This research was led by Marko Prokic of the Department of Physiology at the University of Belgrade.

9. At the time it was last assessed, the goliath frog was listed as endangered because the number of mature individuals was thought to have declined by more than 50 percent over the previous fifteen years.

10. The care and feeding of animals, as described in scientific reports like this, is always fascinating. "For more than a year we have fed them only with finely minced beef liver," the authors wrote in "The aging of *Xenopus laevis*, a South African frog" in the journal *Gerontologia* in 1961.

11. Yes, I have strange hobbies. I started doing this in 2018 on a program called *UnDisciplined* on Utah Public Radio.

12. Writing in the journal *Cells Tissues Organs* in 2018, researchers from the Institute of Zoology in Stuttgart, Germany, called frogs "an undervalued model organism," and argued that frogs are "a more time- and cost-efficient animal model to study human disease alleles and mechanisms."

13. Yes, it's as adorable as it sounds. And yes, we'll be talking more about it—in chapter two.

14. Severe penicillin allergies are rare, with the estimated frequency of anaphylaxis at no more than 5 per 10,000, according to Dr. Sanjib Bhattacharya in "The facts about penicillin allergy: A review," published in the *Journal of Advanced Pharmaceutical Technology & Research* in 2010. That's of no comfort, of course, if you are one of the rare afflicted individuals, which is why individual genetic sequencing is so important to the future of medicine.

15. A lot more people have died from not getting penicillin than from getting it. In "What if Fleming had not discovered penicillin?" a team of researchers from King Saud University and the University of Sheffield imagined a world in which the antibiotic age had never occurred. It was not a pretty picture.

16. In the spring of 2017, for example, the *New York Times* commissioned a survey of more than 1,700 Americans who were asked to identify North Korea on a map. Only about a third could, and the others' guesses, mapped with blue dots on a gray map of Asia, were all over the place, with concentrations of dots in India, Afghanistan, Mongolia, and Vietnam. Those who didn't know were nearly 20 percentage points more likely to believe the United States should engage in a war with that nation. Familiarity may not always breed affinity, but ignorance does seem to correspond to contempt. And geospatial illiteracy is no different than scientific illiteracy—both afflictions lead people to care less about important things.

17. The quest to answer the question "How fast is the universe expanding?" was detailed in Marina Koren's 2017 *Atlantic* article of that name. (Our universe is expanding at a rate of 45 miles per second per megaparsec, according to some estimates.)

18. As Bill Nye famously observed in the first episode of his show *The Eyes of Nye* in 2005.

19. The kindest compliment I've ever received came from the author, essayist, poet, and journalist Sarah Gailey. "You write about terrible things, beautifully," she told me. I certainly aspire to. Many of the pieces referenced here, and other journalistic works from my career as a newspaper reporter and freelance journalist, can be read at mdlaplante.com.

20. Forty-eight hundred pounds, as of the day I wrote this, give or take her last meal.

CHAPTER I: ALL THINGS GREAT AND TALL

1. Delelegn said he thought the society's estimate, published in the journal *Tropical Zoology* in 2000, was high. He felt like the number was likely around 200, noting that he and the others who did the count "knew the elephants and knew where to find them."

2. It was not until 2010 that a genetic study revealed the existence of a third species—the African forest elephant, *Loxodonta cyclotis*.

3. Hybridization of African and Asian elephants is possible in theory, as they share the same number of chromosomes, but there's no clear scientific or conservationist purpose for such a creature, and only one such animal is known to have ever existed. "Motty," who had an Asian mother and an African father, died twelve days after he was born in England's Chester Zoo.

4. Shoshani was killed in a bus bombing in Addis Ababa in 2008. We lost a passionate advocate for elephant research.

5. *Elephants* is one of the best coffee-table books you'll ever find—and you can pick up a used copy for $10 online.

6. In "The sound of silence," published by *Wired* in 2006, the journalist John Geirland wrote about the fascinating history of infrasound, including an exploding-rocket experiment that will "generate data that promises to make the science of infrasound more precise."

7. Though it's not actually explicit in any of Cope's more than 1,300 publications.

8. This was more explicitly stated in 1948 by the German biologist Bernhard Rensch in "Historical changes correlated with evolutionary changes of body size," in the journal *Evolution*.

9. Maureen O'Leary from Stony Brook University led a team of dozens of scientists who "discovered" this hypothetical "mama to us all." Their findings were detailed in "The placental mammal ancestor and the post-K-Pg radiation of placentals" in *Science* in 2008.

10. You might think that we'd have a good idea of the number of animals we share this rock with. As late as 2011, though, a research team from Dalhousie University in Canada noted in a *PLOS Biology* article titled "How many species are there on Earth and in the ocean?" that "the answer to this question remains enigmatic."

11. Bonner's work in books like *Why Size Matters* was often lauded for being "readable" and "unusually clearly written"—which should really be the rule, not the exception, for scientific books.

12. In "Cope's Rule and the dynamics of body mass evolution in North American fossil mammals," published in 1998 by *Science*, Smithsonian paleontologist John Alroy wrote that the body mass estimates for 1,534 North American mammal species showed that new species are 9 percent larger on average than older species in the same genera, and that the effect is stronger for larger animals.

13. "Sometimes, in evolution," journalist Mark Strauss wrote in *National Geographic* in 2016, "the bigger they are, the harder they fall." He was writing about a study, by Hervé Bocheren and his team from Germany's University of Tübingen, that theorized the large size of *Gigantopithecus*, combined with its restricted dietary niche, likely led to its demise at the beginning of the last ice age when tropical forests turned into savannas; the study was published in *Quaternary International* in 2017.

14. Figuratively speaking, of course. Redwoods don't have feet (they have relatively shallow root systems that extend out for hundreds of feet), and neither do whales (although their ancestors appear to have retained legs for millions of years after leaving the land for a life at sea).

15. When the sauropsids and synapsids parted ways, the former ultimately evolved into birds and reptiles, while the latter evolved into mammals—and all of this happened before plants had evolved to produce flowers, Michael Marshall wrote in "Timeline: The evolution of life" in *New Scientist* in 2009.

16. Goodman was prolific, working right up to his death in 2010. He left behind a briefcase of unfinished manuscripts. Years later, papers were still being published under his name. Among his final papers was "Phylogenomic analyses reveal convergent patterns of adaptive evolution in elephant and human ancestries," published in *Proceedings of the National Academy of Sciences* in 2009.

17. I'm not bad at reading scientific papers, but Goodman's report, which is a deep dive into nonsynonymous and synonymous nucleotide substitution rates, would have read like Klingon to me if not for the reporting of science writer extraordinaire Ed Yong, perhaps the best in the game at translating science for "civilians." He wrote "Elephants and humans evolved similar solutions to problems of gas-guzzling brains" for *Discover* in 2009.

18. "Patients and doctors know all too well that cancer is not one disease and there is no singular cure for the complex group of disorders," Dina Fine Maron wrote in "Can we truly cure cancer?" for *Scientific American* in 2016.

19. In preparation for human clinical tests, Alex Stuckey wrote for the *Salt Lake Tribune* in 2017, "Schiffman has formed a spinoff company, PEEL Therapeutics. Peel is the Hebrew word for elephant."

20. Kinzley and her staff were also hesitant to let Osh roam free among the cows since bulls can be very aggressive. One of the Oakland Zoo's female elephants, Medunda, is particularly nervous around Osh. She maintains a friendly demeanor when there is a fence to keep them separate, but will run away if the gate is opened. Zoo officials said Medunda had been attacked by a bull in musth years ago, and had been nervous around bulls since—the connection to the long-term impact of sexual trauma experienced by humans is clear and ripe for further study.

21. Asian elephant spermatozoa, it turns out, don't respond well to cryopreservation, so every new discovery helps, Kiso and fellow researchers reported in "Pretreatment of Asian elephant (*Elephas maximus*) spermatozoa with cholesterol-loaded cyclodextrins and glycerol addition at 4°C improves cryosurvival," published in *Reproduction, Fertility, and Development* in 2015.

22. Among other things, this study on elephant sperm helped inform the work of scientists working to save a rare breed of wild horses in India, Kiso's team reported in "Asian elephant (*Elephas maximus*) seminal plasma correlates with semen quality," which was published by *PLOS One* in 2013.

23. In "Liquid semen storage in elephants (*Elephas maximus* and *Loxodonta africana*): species differences and storage optimization," published in the *Journal of Andrology* in 2011, Kiso and her colleagues demonstrated sperm sensitivity to different "semen extenders" showing different products worked differently on Asian and African elephants.

24. "It is essential," Comizzoli wrote in *Reproduction, Fertility, and Development* in 2013, "that more fundamental studies be directed at more species." In the *Asian Journal of Andrology*, two years later, Comizzoli also noted that biobanking efforts needed to expand "beyond mammalian species, which will offer knowledge and tools to better manage species that serve as valuable biomedical models or require assistance to reverse endangerment."

25. Lamarck wasn't even the first to suggest that acquired characteristics could be inherited; that idea had been around for hundreds if not thousands of years. Scientific historian Michael T. Ghiselin calls the Lamarck of our textbooks "an invention." His essay on that subject, published in 1994 in *The Textbook Letter*, was entitled "The imaginary Lamarck: A look at bogus ‹history› in schoolbooks."

26. There has been a resurgence of appreciation of some Lamarckian principles as we've discovered more about epigenetic inheritance but, as evolutionary biologist David Penny points out in "Epigenetics is a normal science, but don't call it Lamarckian," in the *Journal of Clinical Epigenetics* in 2015, it's not the same thing.

27. Journalist Henry Nicholls detailed the evolution of this idea about evolution in "Giraffes may not have evolved long necks to reach tall trees" for the BBC in 2016.

28. Isbell made this observation along with Truman Young in "Sex differences in giraffe feeding ecology: energetic and social constraints," which was published in the journal *Ethology* in 1991.

29. The publication of "Winning by a neck: sexual selection in the evolution of giraffe" in *The American Naturalist* in 1996 marked a substantial shift in evolutionary thought about how the world's tallest land animal got its long neck.

30. Douglas Cavener led the team that compared giraffe and okapi genome sequences for the paper "Giraffe genome sequence reveals clues to its unique morphology and physiology," published in *Nature Communications* in 2016. Among other observations, the team noted unique genetic changes in signaling pathways that regulate both skeletal and cardiovascular development, and genes that impact mitochondrial metabolism and volatile fatty acids transport, which might have something to do with the ability of giraffes to eat plants that are toxic to other animals.

31. Traditionally, one giraffe species and up to eleven subspecies have been recognized, Janke and his team wrote in "Multi-locus analyses reveal four giraffe species instead of one" in *Current Biology* in 2016, but "their genetic complexity has been underestimated, highlighting the need for greater conservation efforts for the world's tallest mammal."

32. "The whole endeavor of trying to understand and communicate about the diversity of life is being compromised by a naming system that is outdated and has bad consequences," Yale University evolutionary biologist Michael Donoghue complained to *New Scientist* reporter Bob Holmes in 2004.

33. Kevin de Queiroz described some of the complexities around deciding whether something is a species in "Ernst Mayr and the modern concept of species" in the *Proceedings of the National Academy of Sciences* in 2005.

34. Science moves slowly. The 2016 assessment was based upon an interim consensus that a single species of giraffes is resident on the African continent. "Until an extensive reassessment of the taxonomic status of giraffes is completed," the IUCN has declared, "it is premature to alter the taxonomic status quo."

35. There are some species and subspecies that have come close, though, and they may not end up surviving. Western Pacific gray whales are in dire trouble, for instance. The Chinese river dolphin, a toothed whale also known as the baji, was declared functionally extinct in 2006—although there may have been a sighting of one in 2016, according to *Guardian* reporter Tom Phillips in "China's 'extinct' dolphin may have returned to Yangtze river, say conservationists."

36. It's also where I earned my undergraduate degree. Go Beavers.

37. NPR's immersive geology lesson, "Stand at the edge of geologic time," narrated by Kirby, is simply riveting. It was published by NPR in 2016.

38. Whales are still being hit by ships way too often, according to research from the University of Washington that was published in *Marine Mammal Science* in 2014—although the paper, titled "Do ship strikes threaten the recovery of endangered eastern North Pacific blue whales?" answered its titular question with a cautiously optimistic "no."

39. In 2018, Torres and her collaborators published their findings in "Documentation of a New Zealand blue whale population based on multiple lines of evidence" in the journal *Endangered Species Research.*

40. We tend to think of whale migrations as north-to-south affairs, but they also travel latitudinally. Marine biologist Juan Pablo Torres-Florez has identified one female blue that made a record-setting 3,200-mile journey from the Galapagos Islands to the coastal waters of Chile. He nicknamed the global traveler "Isabela," after his child, because, he told me, "I'd like my daughter to know she can do amazing things." Torres-Florez's study, "First documented migratory destination for Eastern South Pacific blue whales," was published in *Marine Mammal Science* in 2015.

41. This is one video of one whale in one place doing one thing—a single-point observation. It's hard to overstate, however, what a truly "aha!" moment this footage offers.

I've shown it to dozens of people, and every one of them sees the same thing: a whale choosing to forgo a small meal because it's a small meal.

42. Intent listening to music in a car, for instance, has been shown to be a significant risk factor for young drivers. It's probably a risk factor for older drivers, too, but only young ones were assessed in the study, "Background music as a risk factor for distraction among young-novice drivers," which was published in the journal *Accident Analysis & Prevention* in 2013.

43. That, it turns out, makes some of the world's largest animals quite similar to some of the smallest. Ants do this as well, as we will examine in chapter eight.

44. We'll dive deeper on that matter when we look at some of the world's smartest animals.

45. I was on my way to work at the McMinnville *News-Register*, "that September day," when I heard the news on my car radio. I had recently begun work as a sportswriter at that small-town Oregon newspaper. It's a reflection of how little we initially knew about the extent and gravity of the attacks that morning, that the first thing my editor said, when I walked in the newsroom door, was, "There is a problem with your column about the cheerleading team; go fix that and then you can work on this New York thing."

46. If that happened suddenly, NASA's Sten Odelwald has observed, the atmosphere would continue to spin at the planet's previous rate of motion, and "all of the land masses would be scoured clean of anything not attached to bedrock. This means rocks, topsoil, trees, buildings, your pet dog, and so on, would be swept away into the atmosphere."

47. Cornelia Dean, who writes beautifully about science, explained this study in "From 9/11, a lesson on whales, noise and stress" for the *New York Times* in 2012.

48. *Los Angeles Times* columnist Michael Hiltzik noted in 2015 four landmark scientific achievements over the prior year—including the first spacecraft landing on a comet, the discovery of the Higgs boson particle, the development of the world's fastest supercomputer, and new research in plant biology. "Then comes the punchline," Hiltzik wrote. "None of these were U.S.-led achievements." Why? Declining investment.

49. "For life to survive as we know it," Marshall told the *New York Times* in 2002, "millions of people are going to have to die. It's sad to say that, but it's true. Millions of people are already dying—it's just gonna have to start happening here." Marshall's logic—that it would take many deaths in the developed world to get people to start caring about the environment—was shocking to me when I first heard it. I've since heard only slightly less violent echoes of this reasoning from a number of respected scientists who have given up hope that anything but a global cataclysm will shake us from our stupor. None of them have ever advocated for terrorist acts, though.

50. And the number of trees has gone down considerably in the past few thousand years. Based on projected tree densities, the authors of "Mapping tree density at a global scale" wrote in *Nature* in 2015, "we estimate that over 15 billion trees are cut down each year, and the global number of trees has fallen by approximately 46% since the start of human civilization."

51. "The story of the carbon is huge," Humboldt State University scientist Robert Van Pelt told *San Jose Mercury News* journalist Paul Rogers in 2016. "The carbon part of a redwood may be more important than the lumber part in the coming decades."

52. In the California Gold Rush of 1849, according to the Save the Redwoods League, these giant trees were logged extensively. Today, only 5 percent of the original forest remains, along a 450-mile coastal strip.

53. I wrote about the great clone with an undergraduate student named Paul Christiansen in a story called "Devastated" for *City Weekly* in 2013. That piece was later honored with a Kavli Award for science journalism by the American Association for the Advancement of Science.

CHAPTER II: ALL THE SMALL THINGS

1. With great humility, van Leeuwenhoek was always writing to other scientists asking for advice, feedback, and even criticism. "I shall be obliged to you for sending me the objections raised to my observations, that I may forward my further observations to your address," he wrote to the German natural philosopher Henry Oldenburg, the father of scientific peer review. "But please to remember who I am and to take my opinions for what they are worth. I have always intended to stand by my speculations and my considerations till I should be better instructed or more experienced and then to abandon my previous opinions and to accept my latest views and put them down in writing."

2. If you are a parent, or plan to become one, I strongly recommend Gilbert and Knight's book, *Dirt Is Good.*

3. Nobody in the world is better at getting researchers to talk like real folk than science journalist Carrie Arnold, who wrote about "The never-ending quest to rewrite the tree of life" for PBS's *NOVA Next* in 2017.

4. In Genesis 1:26, God said, "Let Us make man in Our image, according to Our likeness; and let them rule over the fish of the sea and over the birds of the sky and over the cattle and over all the earth, and over every creeping thing that creeps on the earth." You can blame humans for screwing up nature all you want, but if you believe in this particular god, it sure seems like it's their fault for telling us we were in charge to begin with.

5. The authors of "Scaling laws predict global microbial diversity," which was published in the *Proceedings of the National Academy of Sciences* in 2016, later told me that they might have guessed high, noting that they had begun to wonder, after the paper was published, whether there had in fact been enough time in the 3.5-billion-year history of life on Earth for one trillion species to evolve. Whatever the number, the vastness of microbial diversity is hard to fathom.

6. This approach was also used in a 2002 rape case, also in Washington, resulting in the 2007 arrest of the murderer of a twelve-year-old girl, according to a 2015 article by Sara Jean Green in the *Seattle Times.*

7. The team found twenty genes with significant effects on facial features; the combination of these genes can be applied to approximate the appearance of a face from genetic markers, as described in "Modeling 3D facial shape from DNA" in *PLOS Genetics* in 2014.

8. How good of a picture can we really get from 90 percent of something? Imagine a tyrannosaur chasing after another dinosaur. Then recognize that the picture in your head—hungry eyes, mottled skin, gnashing teeth, lumbering legs, and tiny swaying arms—comes from a species from which we've recovered just a handful of fossils more than 50 percent complete, and just one that was 90 percent intact. (That fossil, known as Sue, is named for adventurer Susan Hendrickson, a salvage-diving, amber-mining, dino-digging legend.) Just like we'll never have a perfect picture of a *T. rex*, it's possible we'll never have a complete image of all of the organisms discovered in Rifle. In both cases, though, there's still a lot we can know from what we have found.

9. Banfield and her team detailed their findings about "ultra-small bacteria" in an article in *Nature Communications* in 2015.

10. That was the case in 2015, when the Rifle study was published, and still in 2018 when I was researching this book.

11. More or less, at least. In "Peculiar properties of mycoplasmas: the smallest self-replicating prokaryotes," published in 1992 in *FEMS Microbiology Letters*, they were called "the smallest self-replicating prokaryotes."

12. "Cum rerum natura nusquam magis quam in minimis tota sit."

13. Globally collaborative science is some of the best science. There were sixty-nine named co-authors who contributed to "The ecoresponsive genome of *Daphnia pulex*" for the journal *Science* in 2011.

14. "Of all the invertebrate genomes sequenced so far," journalist Rebecca Boyle wrote in *Popular Science* in 2011, "the water flea shares the most with us, and scientists hope these shared genes can help them understand how humans respond to environmental threats."

15. Groups like ostracod crustaceans, for instance, as demonstrated by researchers at Canada's Bedford Institute of Oceanography, who published "The genome sizes of ostracod crustaceans correlate with body size" in 2017 in *The Journal of Heredity*.

16. Bristol-based science writer Hayley Bennett writes beautiful British-sounding sentences like "Actually making it, however, is rather trickier." Her 2010 article on "The first synthetic cell" in *Chemistry World* is a great primer on the world's first human-made microbe, a fully functional organism even capable of reproduction.

17. Which is not, in fact, a line from Mary Shelley's *Frankenstein*.

18. Science reporter Rachel Feltman explained this process in accessible detail in "This man-made cell has the smallest genome ever—but a third of its genes are a mystery" for the *Washington Post* in 2016.

19. Because one incredibly complex video is apparently just not enough, OK Go shot two vids for this song. The decision to make a second video seems to have come out of the band's dispute with its record label over how to make viral videos sharable online.

20. Self-proclaimed "professional wildman" Wes Siler appears to have been the first person to notice that OK Go seems to have a strange attraction to the souped-up Escort.

LeMons is a race series in which teams enter junk cars—*lemons*, as it were—that cost no more than $500.

21. Remarkably, a team of scientists from the University of Washington and Microsoft later encoded and decoded that video on strands of DNA, reasoning that it was a delightful representation of how our genome often works.

22. Some historians have speculated that van Leeuwenhoek, who was born in the same year and lived just a few hundred feet from Vermeer, may have served as a model for the painter. The theory has been most notably put forward by author Laura Snyder, who suggests in *Eye of the Beholder* that van Leeuwenhoek might have been the model for Vermeer's *Geographer*.

23. Falkowski's book, *Life's Engines*, gave me a far better appreciation of the role that microbes played in creating the world we know today.

24. Rob Knight's TED Talk is 17 minutes and 20 seconds of whoa. Watch it; it will impact the way you see almost everything.

25. Last time I checked, it was up to 11,190 contributors. Full disclosure: I am one of them. Fuller disclosure: As of this writing, my test kit is still sitting unused on my desk. Because, um, gross.

26. She once told the food science writer Michael Pollan that she was happy to finally be down to giving Knight just one sample a week, but "I keep a couple swabs in my bag at all times," she said in Pollan's May 19, 2013, piece in the *New York Times Magazine*, "because you never know."

27. The really informative—and really funny—*Follow Your Gut* was co-written with excellent science writer Brendan Buhler.

28. American adults with allergies, especially to nuts and seasonal pollen, have low diversity, reduced Clostridiales, and increased Bacteroidales in their gut microbiota, according to «Analysis of the American Gut Project,» published in *EBioMedicine* in 2016.

29. Interestingly, the authors of "Migraines are correlated with higher levels of nitrate-, nitrite-, and nitric oxide-reducing oral microbes in the American Gut Project cohort," published in *mSystems* in 2016, found higher concentrations in oral samples and slight but significant concentrations in fecal samples.

30. "Adults born by Cesarean section appear to have a distinctly different composition of their fecal microbial population," the authors of "Diversity and composition of the adult fecal microbiome associated with history of Cesarean birth or appendectomy: analysis of the American Gut Project" wrote in *EBioMedicine* in 2014. "Whether this distinction was acquired during birth, and whether it affects risk of disease during adulthood, are unknown."

31. Obituaries are amazing things. This one read, in part: "We remember Burt in the special experiences and memories he gave us: Camp Filibert Roth pine cone and water fights, hot spiced cider, apples, speed-driving and hiking, Halloween exam hi-jinks, ecosystems of mystery, creative campfire songs, epic pot lucks, chile-plastic/recycled silverware-jam session, mystery plants, impersonations, 100 station tests, jug band, Paul Bunyan Dance."

32. "As far as knowledge of Hawai'ian insects, he was the best," Dick Tsuda, a fellow entomology researcher who was one of Beardsley's students, told the newspaper.

33. *Kikiki huna's* discovery, described in 2000 in "A new genus of fairyfly, Kikiki, from the Hawaiian Islands," came while researchers were engaged in a generic review of the Hawai'ian genera of Mymaridae. Sometimes the most amazing things in the world are discovered when we're not really looking.

34. Hat tip to my favorite beta reader, Sarah Gailey . . . I wasn't clever enough to identify this nomenclatural Easter egg.

35. In this same study, an even smaller size for *K. huna* was identified, and another tiny insect was discovered as well. "A new genus and species of fairyfly, *Tinkerbella nana* (Hymenoptera, Mymaridae), with comments on its sister genus *Kikiki*, and discussion on small size limits in arthropods," was published in the *Journal of Hymenoptera Research* in 2013.

36. As of 2009, the costs of brown tree snake detection and control were estimated at $2.5 million per year, according to the US State Department.

37. The authors of "Effects of an invasive predator cascade to plants via mutualism disruption," published in *Nature* in 2017, conservatively estimated the loss could be "only" 61 percent, but might be as high as 92 percent. This is terrifying.

38. Just ten years after its arrival in the United States, the tiger mosquito had spread to 678 counties, according to the 1997 report "*Aedes albopictus* in the United States: ten-year presence and public health implications" in *Emerging Infectious Diseases.*

39. Not to be mistaken for the also-invasive zebra mussel, though the quagga is named, as it happens, for an extinct species of zebra.

40. This observation comes by way of *New York Times* science writer Erica Goode in "Invasive species aren't always unwanted." Goode is one of the best in the field at putting the march of science into social context.

41. Ken Thompson, a biological devil's advocate from the University of Sheffield, wrote about Elton in his brilliant book *Where Do Camels Belong?*

42. A good background on the history of "invasive" things is in Charles Elton's book, The *Ecology of Invasions by Animals and Plants.*

43. This in a year in which the Soviet Union tested at least thirty-six nuclear bombs, and the United States and United Kingdom entered a nuclear mutual defense agreement.

44. The website Untamed Science offers a great video on the nuances of this controversy. "Invasive species–Fight 'em or throw in the towel?" is worth a watch.

45. That contention was made on the National Wildlife Federation's website in 2013. That page has since gone dormant, but it's still available via archive.org.

46. "Today's management approaches must recognize that the natural systems of the past are changing forever," Davis and his colleagues wrote in "Don't judge species by their origins."

47. A thin plastic disc upon which a tight spiral of grooves are read by a needle, creating sound.

48. Their study, complete with really cool photos and a map of where they found the little frogs, was published in *PLOS One* in 2012.

49. This may lead to first-line antiviral treatments during influenza outbreaks, according to

"An amphibian host defense peptide is virucidal for human H1 hemagglutinin-bearing influenza viruses," published in the journal *Immunity* in 2017.

50. The common prefix for both of these micro-vertebrates, "paedo," means "child" or "childlike."

51. At that point, the Thai population of bumblebee bats had been impacted by collectors and tourists, such that it was getting harder to study these animals in anything close to their natural state. The Myanmar find offered a second chance for scientists to see this species in the conditions in which it evolved.

52. According to "The evolution of sensory divergence in the context of limited gene flow in the bumblebee bat," published in *Nature Communications* in 2011, geographic distance has a larger role in limiting gene flow than echolocation divergence.

53. When biologists Ahmet Selcuk and Haluk Kefelioglu took a survey of tiny mammals in northern Turkey, they had to search through 100 pellets to find just one with an Etruscan shrew's remains in it. Their study, "New record of *Suncus etruscus Savi*, 1822 (Mammalia: Soricomorpha) in Northern Turkey," published in *Biharean Biologist* in 2016, revealed the first record of *S. etruscus* in northeastern Anatolia, Turkey.

54. Despite having a brain that is only 64 mg in weight, the Etruscan shrew "demonstrates a wide range of social and exploratory behaviors, as well as sophisticated prey-capture capabilities and unique adaptations of the cardiovascular and respiratory systems to small body size," according to Robert Naumann in "Even the smallest mammalian brain has yet to reveal its secrets," published in *Highlights and Perspectives on Evolutionary Neuroscience* in 2015.

55. "Small size, high-speed behavior, and extreme dependence on touch are not coincidental, but reflect an evolutionary strategy, in which the metabolic costs of small body size are outweighed by the advantages of being a short-range high-speed touch and kill predator," the researchers wrote in 2011 in "The neurobiology of Etruscan shrew active touch," in *Philosophical Transactions of the Royal Society B: Biological Sciences.*

56. "One of the beauties of whiskers is that the whisker can get damaged but still works—all the sensing is done at the root," Alan Winfield of the Bristol Robotics Laboratory told CNN in 2013.

CHAPTER III: THE OLD DOMINION

1. The Smithsonian's spectacular "Human Evolution Timeline Interactive" exhibit notes that some of the most important human milestones—brain development, the control of fire, and bipedalism—have occurred during times of the greatest climate fluctuation.

2. I've heard scientists throw around estimates of 30,000 years, too. But everyone concedes that these are guesses.

3. That's the tallest and shortest averages from a group of 98 nations, although Africa was largely left out of the list these are drawn from, published by the *Telegraph* in 2017.

4. The institution has a super-cool acronym: RED Lab.

5. Pollen in sediment offers another way to see the world thousands of years before the common era, as highlighted in 2017 in the paper "Climate variability and fire effects on quaking aspen in the central Rocky Mountains, USA" in the *Journal of Biogeography*.

6. That is one of the key takeaways of *Inheritance: How Our Genes Change Our Lives and Our Lives Change Our Genes,* which I co-wrote with geneticist Sharon Moalem in 2015. It reveals how genetic breakthroughs are completely transforming our understanding of both the world and our lives.

7. So protective is the Tasmanian government of *L. tasmanica* that when the journalist Graham Lloyd set out to write about it a few years ago, he had to sign a legally binding agreement not to disclose its location, and the Tasmanian National Parks rangers who were his guides still decided to blindfold him, just to make sure. Lloyd's article was published in *The Australian* in 2014.

8. With infinite time and infinite pianos, an infinite number of monkeys would eventually perform one of Mozart's concertos, and enough mules have had enough sex to have a foal or two (though when it happened in Morocco in 2002, locals reportedly feared it signaled the end of the world, as Nancy Lofholm reported for the *Denver Post* in 2007). This could potentially happen with triploid plants, too. Sexual reproduction, that is. Not the end of the world.

9. Two other known *G. renwickiana* clones are about half the size of the biggest. That's still *really* big. All of the remaining known clones were mapped in "Spatial genetic structure reflects extensive clonality, low genotypic diversity and habitat fragmentation in *Grevillea renwickiana* (Proteaceae), a rare, sterile shrub from south-eastern Australia," published in *Annals of Botany* in 2014.

10. Maynard Smith observed in 1958 that *Drosophila subobscura* flies live longer when they are sterile, a finding that was later replicated in the more popular lab species, *D. melanogaster*, as well as in *Caenorhabditis elegans*, the extensively studied roundworm that Cynthia Keaton used to discover genetic causes for aging. A good primer on these studies can be found in "Survival costs of reproduction in *Drosophila*" by Thomas Flatt in the May 2011 edition of *Experimental Gerontology*.

11. Their study, "Human longevity at the cost of reproductive success: evidence from global data" was published in the *Journal of Evolutionary Biology* in 2001.

12. Rogers (who is not the same Paul Rogers as the California journalist who wrote about how redwoods sequester carbon) sounded the alarm with Darren McAvoy in "Mule deer impede Pando's recovery: Implications for aspen resilience from a single-genotype forest," which was published in *PLOS One* in 2018.

13. This was game management at its worst. The Mountain Lion Foundation has published a thorough timeline of the various bounty programs that were in place across North America, starting in the 1500s, at mountainlion.org.

14. Science writer Kate Tobin did a marvelous job of bringing this slow-moving ecological story to life in "Did wolves help restore trees to Yellowstone?" for PBS in 2015.

15. "We can place this book on the shelf that holds the writings of Thoreau and John Muir," the *San Francisco Chronicle* once wrote of *A Sand County Almanac*. It's on my desk next to the Bible, and I reach for Leopold's words far more often.

16. "The story, which quickly grew as twisted as a bristlecone's, went something like this: Ancient tree grows for millennia on a Nevada mountaintop until it becomes the World's Oldest Tree," Shaun McKinnon wrote in the *Arizona Republic* in 2015. "Because it's so remote, hardly anyone ever sees it or thinks of it."

17. It will likely come as no surprise to you, at this point, to learn that Harlan's lengthy obituary in the *Arizona Daily Star* did not include the words "claimed to have identified the oldest known bristlecone pine in the world."

18. Sarah Zielinski wrote about Procopius and new analyses that combine ice cores with data from tree rings in "Sixth-century mystery tied to not one, but two, volcanic eruptions," for Smithsonian.com in 2015.

19. Salzer and research partner Malcom Hughes connected frost rings to volcanic eruptions over a 5,000-year period in "Bristlecone pine tree rings and volcanic eruptions over the last 5000 years" for the journal *Quaternary Research* in 2006.

20. Another Salzer study, "Recent unprecedented tree-ring growth in bristlecone pine at the highest elevations and possible causes," published in *Proceedings of the National Academy of Sciences* in 2009, offered this superlative observation.

21. The results from that study, titled "Analysis of telomere length and telomerase activity in tree species of various lifespans, and with age in the bristlecone pine *Pinus longaeva*" and published in the journal *Rejuvenation Research* in 2006, suggested that increased telomere length may contribute to the longevity of bristlecones.

22. "We found no evidence of mutational aging," the authors of "Does bristlecone pine senesce?" wrote in 2001 in *Experimental Gerontology*. "We conclude that the concept of senescence does not apply to these trees."

23. "At first sight negligible aging, like cessation of aging, does not seem to fit neatly under Hamilton's theory of a decline of the force of natural selection with increasing age," Martínez wrote in *Frontiers in Genetics*, alluding to William Hamilton's 1966 paper, which sought to explain aging and supplied basic lifetime scaling forces for natural selection.

24. While one species of hydra, *Hydra vulgaris,* has shown no signs of aging, another, *Hydra oligactis,* can be induced to undergo reproduction-associated senescence, according to «*Hydra* as a tractable, long-lived model system for senescence,» published in *Invertebrate Reproduction & Development* in 2015.

25. Davy's collaborators were researchers from the Honolulu-Asia Aging Study, a longitudinal investigation of the rates, risk factors, and genetic causes of cognitive decline and dementia in aged Japanese American men. "Exceptional longevity: insights from hydra to humans" was published in *Current Topics in Developmental Biology.*

26. "Constant mortality and fertility over age in Hydra," published in *Proceedings of the National Academy of Sciences* in 2015, looked at 2,256 hydra in two laboratories, with cohort ages up to forty years.

27. There are several versions of this story. Michael Cohen's 1998 book, *A Garden of Bristlecones: Tales of Change in the Great Basin,* includes what appears to be the most well-sourced version.

28. The episode featuring this story was broadcast on December 11, 2001.

29. The Currey tale is the subject of one of the best *Radiolab* episodes ever. And that's saying something, because *Radiolab* is simply brilliant.

30. By comparing the rings of living (or recently killed) clams to shells from dead clams, researchers have been able to build a timeline of abut 1,000 years that offers information about temperature and salinity of the seawater at the time of the growth of each ring. Researchers from Cardiff University and Bangor University explained this in "What 500-year-old clams can tell us about climate change" in *The Conversation* in 2016.

31. "It's an epic shellfish saga," science reporter Rebecca Morelle wrote for the BBC, "with all the makings of a rather tasty story."

32. This is an absolutely terrifying study, published in 2016 in *Nature Communications* and sort of hidden under the inconspicuous title "Annually resolved North Atlantic marine climate over the last millennium." In short, it says that a geologic effect that appears to be as old as the oceans themselves has changed—because of us.

33. The process used to arrive at this superlative conclusion is explained in "Siliceous deep-sea sponge *Monorhaphis chuni*: A potential paleoclimate archive in ancient animals," published in 2012 by the journal *Chemical Geology*.

34. And we know what the genes are that create the stem cells, thanks to Werner Müller's "The stem cell concept in sponges (Porifera): metazoan traits," published in 2006 by *Seminars in Cell and Developmental Biology*.

35. "Like in chess, in which all pieces have to protect the king, the condition for a long life in a perennial plant is to protect the roots (or at least, the capability to regenerate roots quickly . . .)," Munné-Bosch wrote in *Plant Physiology* in 2014.

36. They also found changes to genes for thermoregulation, sensory perception, dietary adaptations, and immune response, according to "Evolution of longevity from the bowhead whale genome," published in 2015 by *Cell Reports*.

37. Two studies, one from *Ecology and Evolution* in 2014 ("Stressful environments can indirectly select for increased longevity") and another from *Ageing Research Reviews* in 2015 ("Aging and longevity in the simplest animals and the quest for immortality"), help build the case for stress and simplicity.

38. Although China didn't start issuing formal birth certificates until long after the Bapan centenarians were born, family documents, military records, and the culture of *shengxiao*, the twelve-year Chinese zodiac, provides compelling evidence that these individuals are indeed as old as reported.

CHAPTER IV: FAST TIMES

1. The researchers recorded data from 367 hunts. In no case did a cheetah exceed 58 miles per hour, and the runs went for an average distance of just 173 meters, according to "Dynamics and energetics of hunting in the cheetah," a series of research projects supported by the Royal Veterinary College in 2016 and 2017.

2. BBC Earth did a good job of explaining this in "The cheetah: nature's need for speed," for *Mother Jones* in 2011.

3. Veterinary physiologist Naomi Wada's team demonstrated the ways in which cheetahs operate like a rear-wheel-drive car, with the thrust power coming from the back legs and turning and breaking power up front, in "Distribution of muscle fibers in skeletal muscles of the cheetah (*Acinonyx jubatus*)" for *Mammalian Biology* in 2013.

4. Science journalist Matt Bardo explained Wada's work in "Cheetah's speed secrets are revealed" for BBC Nature in 2012.

5. In "A general scaling law reveals why the largest animals are not the fastest," the researchers unpacked a fundamental constraint on the upper limit of animal movement, offering a better understanding speed in nature. Their work was published in *Nature Ecology & Evolution* in 2017.

6. Science reporter Helen Briggs explained "Why the cheetah is a champion sprinter" for BBC News in 2017.

7. "To get around those limitations," journalist Sid Perkins wrote for *Science Magazine* in 2017, "Hirt and her colleagues looked at previously collected data for a wide variety of creatures, including ectotherms (so-called cold-blooded animals) as well as warm-blooded endotherms."

8. The researchers created charts—with little silhouettes of animals, no less—that demonstrate the ways the curves are similar.

9. And the science is . . . well . . . it's just bad, as Phil Plait and Daniel Hubbard explained for *Slate* in 2015. But "nerd sexy" Jeff Goldblum is still dreamy. And dinosaurs eat people. And that's all that matters.

10. But a velociraptor might have done 35!

11. In "Excess of genomic defects in a woolly mammoth on Wrangel Island," researchers wrote about their findings of an excess of deletions, an increase in the proportion of deletions that affected gene sequences, and an excess of truncated, incomplete, or nonfunctional genetic protein products, none of which is good for long-term viability of a population. Their work was published in *PLOS Genetics* in 2017.

12. When "Skin grafts and cheetahs" was published in *Scientific Correspondence* in 1996, a lot of people didn't believe it, but further experiments demonstrated that populations of other animals with low levels of genetic variation can also have significant histocompatibility.

13. According to "Genomic legacy of the African cheetah, *Acinonyx jubatus*," a study to which O'Brien contributed in 2015 for *Genome Biology*, no other mammal we know of comes anywhere close.

14. At least with one population of free-ranging cheetahs in Namibia, however, this does not appear to have affected their immune systems, according to researchers from Germany's Leibniz Institute for Zoo and Wildlife Research, who wrote about the Namibian cheetahs in 2011 in the journal *Molecular Biology and Evolution*.

15. It's one thing to survive one population bottleneck. Surviving two is improbable at best.

16. Studies like "Tracking data from nine free-roaming cheetahs (*Acinonyx jubatus*) collared in the Thabazimbi area, Limpopo Province, South Africa" from the Endangered Wildlife Trust's Carnivore Conservation Program, published in 2017 by *Biodiversity Data Journal*, offer a very depressing glimpse into the lives and deaths of endangered species.

17. Those were Asiatic cheetahs in Iran, which are at even greater risk of disappearing, according to "The critically endangered Asiatic cheetah *Acinonyx jubatus venaticus* in Iran: a review of recent distribution, and conservation status,» from the journal *Biodiversity and Conservation* in 2017.

18. Based on body shape alone, the pronghorn probably shouldn't be able to run as fast and far as it does. Which is why researchers writing for *Nature* in 1991 concluded "their performance is achieved by an extraordinary capacity to consume and process enough oxygen." Their report was titled "Running energetics in the pronghorn antelope."

19. Diving deeper into the linguistic weeds, the word "buffalo" derives from from the Greek "boubalos," which referenced both wild oxen and, as it happens, antelope.

20. University of Chicago Marine Biological Laboratory manager David Remsen wrote in a 2016 paper for the journal *ZooKeys*, called "The use and limits of scientific names in biological informatics," that the consistency in the relationship between names and taxa "places limits on how scientific names may be used in biological informatics in initially anchoring, and in the subsequent retrieval and integration, of relevant biodiversity information."

21. Science writer Brian Switek explained how the pronghorn won the evolutionary arms race, then refused to disarm, for NationalGeographic.com in 2013 in "Did false cheetahs give pronghorn a need for speed?"

22. For all intents and purposes they're running from ghosts, Byers writes in *Pronghorn: Social Adaptations and the Ghosts of Predators Past.*

23. Oh. And don't forget the hyenas that were equipped with cheetah-like limbs and huge jaws. As Byers told *New York Times* reporter Carol Kaesuk Yoon in 1996, "I don't think there's a predator alive today that would've been as ferocious as that long-legged hyena would've been."

24. The story of how scientists concluded that a human ancestor called the Taung Child had been killed by an eagle is a truly fascinating tale. Riaan Wolmarans covered the story for Agence-France-Presse's South African Press Association in 2006 in "Taung child's death puzzle finally solved."

25. In "Animals that are peerless athletes," published in 1993, *Times* reporter Natalie Angier described the pronghorn as "a goat-sized ungulate that may rank as the greatest athlete alive."

26. This has really got to be one of the most kick-ass biology labs in the world.

27. We may have made a titular claim upon it, but *Periplaneta americana* did not, in fact, come from America. It originated in Africa and came to the New World by way of ship in the early 1600s. Cockroaches should be, then, a very small reminder of the abject evil upon which the United States was built. They are part of a legacy of slaveholding in America and, in my mind, are a symbol of the ways in which that legacy still permeates every corner of the nation in which I was born and live.

28. Pretty much anyone could set up an experiment like this. That's part of what I love about it. The methods were described in the University of Florida's Book of Insect Records in 1999.

29. Moreover, they are able to withstand sidewalk surface temperatures from 40 to 60 degrees Celsius. If I were running on a sidewalk that hot with bare feet, I'd probably be pretty fast, too. All of this and more was described in "Exceptional locomotory performance in *Paratarsotomus macropalpis* mites,» published in *The FASEB Journal* in 2014.

30. And, if we're being technical, it actually might have been walking, not running. Runners, after all, leave the ground between strides. Like competitive walkers, who are permitted to only have one leg leave the ground at a time, mites never have more than four of their eight feet off the ground when they're dashing from place to place.

31. In "Putting nanoparticles to work: Self-propelled inorganic micro- and nanomotors," in *Anisotropic and Shape-Selective Nanomaterials,* Kaitlin Coopersmith wrote about "machines inspired by nature's elegant use of chemical gradients and cellular tracks for independently driven molecular processes."

32. The videos included in "Exceptional running and turning performance in a mite" are thrilling to watch. The research was published by the *Journal of Experimental Biology* in 2016.

33. Psychologist Robin Rosenberg's *What's the Matter with Batman? An Unauthorized Clinical Look under the Mask of the Caped Crusader* is fun reading for anyone who wants their hero ruined for good.

34. Tom Harpole's 2015 article for *Air & Space Magazine* on Ken Franklin's quest to understand falcon flight, "Falling with the falcon," is riveting reading.

35. 27 Gs of deceleration, according to Harpole.

36. Wild observations alone likely would never have produced this finding. In freely roaming falcons, the authors of "Diving-flight aerodynamics of a peregrine falcon" wrote in 2014 for *PLOS One*, "these high velocities prohibit a precise determination of flight parameters such as velocity and acceleration as well as body shape and wing contour."

37. Yes, that's seven years before Kitty Hawk. "If it seems there are a few caveats to the Wright brothers' achievement 106 years ago," Jason Paul wrote for *Wired* in 2009, "it's because there had been several people before them who had already managed to get aloft in some sort of device, including the brothers themselves. What the Wrights did was put it all together in a way that made the airplane workable."

38. This was part of the European Commission's PEL-SKIN project, aimed at discovering an airfoil coating to improve the aerodynamic performance of aircraft. It was described in "The PELskin project-part V: towards the control of the flow around aerofoils at high angle of attack using a self-activated deployable flap" for the journal *Meccanica* in 2017.

39. Although the figure was apparently quite pervasive in the pre-internet era as well, once websites began sharing it, and citing each other, it seems it stopped mattering where the original information came from.

40. Like "Hydrodynamic characteristics of the sailfish (*Istiophorus platypterus*) and swordfish (*Xiphias gladius*) in gliding postures at their cruise speeds," published by *PLOS One* in 2013.

41. One possible source is Arthur Upfield's "The Mystery of Swordfish Reef," a fictional story published in 1939 that puts Upfield's protagonist, Inspector Napoleon "Bony" Bonaparte, on the end of a rod and reel being torn of its line at lightning speed. "There had been nine hundred yards wound on it," Upfield wrote. "Now there were only seven hundred yards. Three seconds later there were but six hundred yards." Bony's companions compare him to Zane Grey in that book.

42. "To put this in perspective," oceanographer Molly Lutcavage's Large Pelagics Research Center reported on *Medium* in 2015, the sailfish's 1.79 G acceleration could be compared "to the sports car Bugatti Veyron's 1.55 G, which reaches 0–60 mph in 2.4 seconds" (and the fish would be faster!).

43. That study was a hoot, though. A group of mechanical engineers from Seoul National University caught a seven-foot sailfish in the South China Sea, killed it, froze it, stuffed it, put it into a wind tunnel, and then gave it several nose jobs to see if the size of the fish's bill had an impact on drag. It didn't, according to "Hydrodynamic characteristics of the sailfish (*Istiophorus platypterus*) and swordfish (*Xiphias gladius*) in gliding postures at their cruise speeds," published in 2013 by *PLOS One*.

44. Premium bluefin can go for more than $100,000 at auction, and one 489-pound fish went for $1.7 million in Japan in 2013. It was a bit of a publicity stunt by Tokyo Zanmai restaurant chain at the first auction of the year, where restaurant chains try to outbid one another for the "New Year Tuna." Still, that's about $3,500 a pound, according to Faine Greenwood's piece for PRI's *GlobalPost* in 2013.

45. "Electronic tagging of Atlantic bluefin tuna (*Thunnus thynnus*, L.) reveals habitat use and behaviors in the Mediterranean Sea," for *PLOS One* in 2015, confirmed the Med is used as a spawning ground and overwintering foraging location.

46. Eastern and western Atlantic bluefin differ greatly in size; the former is about 10 times the size of the latter. The genes of different populations of this fish vary widely, too—so much so that when scientists studied nearly 1,000 eggs from Mediterranean bluefin, they found 129 different haplotypes, more than half of which had never been detected in the Med before. "Individual spawning duration of captive Atlantic bluefin tuna (*Thunnus thynnus*) revealed by mitochondrial DNA analysis of eggs" was published in *PLOS One* in 2015.

47. A double-edged sword: The same science that is being used to understand and protect tuna can be used to identify foraging aggregations, leading to concentrated fishing efforts, according to "Seasonal movements, aggregations and diving behavior of Atlantic bluefin tuna (*Thunnus thynnus*) revealed with archival tags," in *PLOS One* in 2009.

48. "Dolphins have been used in mine hunting in the Iraq war and even anti-enemy swimmer roles in Vietnam. And the reality is that for the time being, animals have superior agility and even sensors than most robotic systems," defense strategist P. W. Singer told Carl Prine of the *San Diego Union-Tribune* in 2017. As robots continue to advance in speed, maneuverability, and autonomy, however, this is likely to change.

49. Researchers from the National Marine Mammal Foundation, which works with the US Navy, have published scores of studies in recent years.

50. "If you're going to keep them in captivity then the research you do with them has to have a direct, positive input into their conservation in the wild," Rose told San Diego's *CBS News* affiliate in 2017.

51. Some studies, like "Speed limits on swimming of fishes and cetaceans," published in 2008 in the *Journal of the Royal Society*, have suggested that burst speeds as high as

33 miles per hour are possible for dolphins, but that's still a far cry from what a tuna can do.

CHAPTER V: AURAL SECTS

1. That's apparently $pLW^2/6$, if you're trying this out at home.

2. "Some studies say that women find deeper voices to be more attractive," Knapp told *Salt Lake Tribune* reporter Rich Kane in 2015. "Other studies have found that the deeper a man's voice pitch, the more women they've slept with."

3. The sports car–owning men were six times less likely to say they were "smaller than average" than their partners. The "fast car, small manhood" survey might align with a whole lot of anecdotal evidence, but it absolutely must be taken with a grain of salt. The purported survey's sponsor reported its results, but not its methodology, in the *Daily Mail* in 2014.

4. "The problem for researchers has been that they have had to rely on participants providing their own measurements," Tom Hickman wrote for *Salon* in 2013. "Where men and their penises are concerned there are lies, damned lies, and self measurements."

5. At least until adulthood, when we typically lose the ability to hear frequencies at the top end of that spectrum.

6. For one study, Payne examined 548 songs from humpback whales in the Pacific and Atlantic oceans, finding rhyme-like structures occur "in songs containing the most material to be remembered," according to her 1998 report with Linda Guinee, titled "Rhyme-like repetitions in songs of humpback whales" in the journal *Ethology*.

7. Payne's book, *Silent Thunder: In the Presence of Elephants*, is pure joy.

8. Elephants were the first terrestrial mammals reported to produce infrasound. That revelation was published in "Infrasonic calls of the Asian elephant (*Elephas maximus*)" in *Behavioral Ecology and Sociobiology* in 1986.

9. Famed elephant researcher Joyce Poole's research into elephant musth has been essential for understanding reproduction. Poole's "Rutting behavior in African elephants: the phenomenon of musth" was published in the journal *Behavior* in 1986.

10. It now seems obvious that elephants would communicate in different ways depending on how far away other elephants were, but as late as 2000, when W. R. Langbauer wrote about "Elephant communication" for the journal *Zoo Biology*, it was a revelatory finding.

11. There have been at least 164 different species of proboscideans. Only a few are left, according to "Understanding proboscidean evolution: a formidable task," published in 1998 by *Trends in Ecology & Evolution*.

12. It took an expert in whales to unlock the secret of elephant infrasound. And it took elephant infrasound to help us better understand aquatic audition in marine mammals like whales, like the adaptations described in Darlene Ketten's "The marine mammal ear: specializations for aquatic audition and echolocation" for *The Evolutionary Biology of Hearing* in 1992.

13. Clues offered by one type of animal often help us ask better questions about other animals, even ones with which we think we have great familiarity, like the cows that were the subject of "Vocal behavior in cattle: the animal's commentary on its biological processes and welfare," published in 2000 by *Applied Animal Behavior Science.*

14. Scientists studying "distress vocalizations" in baby animals have relied on what has been learned about how elephant calves communicate. Such research is described in "Do infant rats cry?" published in 2001 in *Psychological Review.*

15. Once we realized that many animals hear outside the frequencies we hear, we realized that what *we* did outside those frequencies, like the "pings" of military and research sonar systems, mattered to many creatures, like those described in "Anatomy and physics of the exceptional sensitivity of dolphin hearing (Odontoceti: Cetacea)," published in 2010 by the *Journal of Comparative Physiology.*

16. Recorded sound analysis can give us "a deeper understanding of their meaning and significance with respect to well-being of farm animals," the authors of "Vocalization of farm animals as a measure of welfare" wrote in *Applied Animal Behavior Science* in 2004.

17. As we've learned more about how rodents communicate distress, we have been able to connect verbalizations to brain activity and other actions, such as those described in "The ascent of mouse: advances in modelling human depression and anxiety" from *Nature Reviews Drug Discovery* in 2005. We couldn't have done this if we hadn't started looking for sounds outside our range of hearing.

18. The process Galton took to develop and test his whistle is an important example of how science needs engineering. A good primer on the history of this invention is "The Galton whistle" in the March 2009 edition of the *Observer,* a publication of the Association for Psychological Science.

19. Carol Kaesuk Yoon wrote an appreciation of Griffin's fascinating life for the *New York Times* in 2003.

20. And according to "Communication of adult rats by ultrasonic vocalization: biological, sociobiological, and neuroscience approaches" from *ILAR Journal* in 2009, these calls are reliable predictors of increased chemical activity in the brain.

21. But not guinea pigs, opossums, and beluga whales, according to *Comparative Hearing: Mammals,* published in 1994.

22. Martin and University of New South Wales professor Tracey Rogers wrote about this in "Why do elephants bellow but whales squeak like a mouse?" for *The Conversation* in 2016.

23. "Does size matter? Examining the drivers of mammalian vocalizations" was published in the journal *Evolution* in 2016.

24. Also known as "rock-paper-scissors," and by a variety of other names, the game may have originated in China during the Han Dynasty.

25. There was absolutely no sense of humor in this study, published in the journal *Applied Entomology and Zoology* in 1990, which was called "Sound production in *Synaptonecta* *issa* (Heteroptera: Corixidae, Micronectinae)—an Asian bug that turned up in a New Zealand aquarium." There should have been.

26. Technically speaking, it's the insect's right penis. Yes, there is a left one, too.

27. A motorcycle at 25 feet comes in at 90 decibels; a power lawn mower registers at 100.

28. The charts published in the 2011 study "So small, so loud: extremely high sound pressure level from a pygmy aquatic insect (Corixidae, Micronectinae)" in *PLOS One* show the water boatman is an exceptional outlier.

29. The fascinating physics of a pistol shrimp's snapping action were described in "Vortex formation with a snapping shrimp claw" for *PLOS One* in 2013.

30. It was a popping bubble, not a snapping claw, that was causing all that racket, according to a 2000 study called "How snapping shrimp snap: through cavitating bubbles" from *Science.*

31. In a moment of if-you-can't-beat-them-join-them brilliance during World War II, the US Navy used the little snappers as sound screens to hide its submarines from hydrophones in Tokyo Bay, according to "Shrimp aided submarines" from the Associated Press on March 16, 1947.

32. When a long-held assumption is destroyed, new research applications are never far behind. That's one of the lessons in the article "Effects of ghost cavitation cloud on near-field hydrophones measurements in the seismic air gun arrays," presented at the European Association of Geoscientists and Engineers' 2017 annual conference.

33. American inventor Amos Dolbear invented a telephone more than a decade before Alexander Graham Bell, but didn't have Bell's command of "patent office formalities," according to an 1881 article in *Scientific American.*

34. Scientists thought they'd re-found it four times between 1891 and 2012. They hadn't. The story was unpacked in "The spider-like katydid *Arachnoscelis* (Orthoptera: Tettigoniidae: Listroscelidinae): anatomical study of the genus," published in *Zootaxa* in 2013.

35. I cannot say enough nice things about this book, which helps fill in the gaps between "life began" and "then dinosaurs arrived."

36. This is called the photoacoustic effect, and best we can tell it was indeed first discovered by Alexander Graham Bell.

37. Phil Senter covered this in a chapter on the pre-Cenozoic evolution of animal acoustic behavior in *Historical Biology* in 2009.

38. The large square-shaped vocal fold in cats of the genus *Panthera* extends the dynamic range of the cats who use it to roar, according to "Adapted to roar: functional morphology of tiger and lion vocal folds" from *PLOS One* in 2011.

39. That's according to "Coos, booms, and hoots: the evolution of closed-mouth vocal behavior in birds," from the journal *Evolution* in 2016.

40. "Changes in acoustic interactions may thus go with the break of maternal care as well as dispersal of juvenile crocodilians," the team wrote in "Size does matter: crocodile mothers react more to the voice of smaller offspring" for the journal *Scientific Reports* in 2015. This trait might have been present in other archosaurs like dinosaurs and pterosaurs.

41. "In Puerto Rico, where the coqui frog has always been part of the island's life, there is a deep respect for it," Will White and Kate Sensenig wrote for the *Washington Post* in 2015. Not so much in Hawai'i.

42. "The frogs' high-pitched nighttime mating calls have caused residents many sleepless nights," Sarah Lin wrote for the *Los Angeles Times* in 2014.

43. That was one of many plans for defeating the coqui and other nonnative species back in 2002, when science writer Janet Raloff noted that, "after working their way through soaps, surfactants, and off-the-shelf pesticides" anti-frog warriors considered "products in the grocery store, including acetaminophen (Tylenol) and cigarette nicotine" before settling on "a caffeine-rich anti-sleep preparation." Raloff's article was published in *Science News* in 2002.

44. The state's Plant Industry Division offered the 10,000-amphibians-per-acre fact as an indicator of the frog's negative impact on the native environment. That might very well be true, but they were putting the cart before the frog.

45. Singer's *Panic in Paradise: Invasive Species, Hysteria and the Hawaiian Coqui Frog War* covers his concerns in depth.

46. Interestingly, when the playback intensity of the two-note call reaches a certain threshold, the males answer by uttering just a "koh" note almost half the time, according to "Communicative significance of the two-note call of the tree frog *Eleutherodactylus coqui*" from the *Journal of Comparative Physiology* in 1978. It seems that when one coqui senses another is nearing, the aggressive part of its call takes center stage.

47. The American soccer player John Harkes could be Exhibit A: Within years of his arrival, the first Yank to play in the English Premier League had swapped out his northern New Jersey accent for one that was decidedly more British—and hints of the English Midlands were still dripping from his lips a decade after he returned to the United States.

48. University of Hawaii at Hilo researchers Francis Benevides and William Mautz examined duration, inter-note interval, call repetition period, center frequency, and bandwidth in "Temporal and spectral characteristics of the male *Eleutherodactylus coqui* two-note vocalization in Hawaii" for the journal *Bioacoustics* in 2013.

49. Narins is a great example of what scientists can do when they are able to focus on something for long periods of time. The study about the two parts of the coqui call was published in 1978; the follow-up came in 2014. It was titled "Climate change and frog calls: long-term correlations along a tropical altitudinal gradient" and published in *Proceedings of the Royal Society B: Biological Sciences.*

50. The study was called "Sound perception by two species of wax moths" and Spangler made no superlative declaration. Thus, when it was published by the relatively obscure *Annals of the Entomological Society of America* in 1983, it was widely missed.

51. "Possibly this moth's success depends in part upon the adaptability and multiple uses of its ear," Spangler observed in 1984 in "Responses of the greater wax moth, *Galleria mellonella* L. (Lepidoptera: Pyralidae) to continuous high-frequency sound" for the *Journal of the Kansas Entomological Society.*

52. "This moth can hear the calls of any bat," James Windmill, an acoustical engineer at the University of Strathclyde, marveled to *Nature* science writer Ed Yong in 2013.

53. Some of Spangler's other work was actually cited in "Extremely high frequency sensitivity in a 'simple' ear" for *Biology Letters* in 2013, but not his work on frequency range.

54. This is one way scientists have concluded that these early bat species echolocated, Michael Novacek reported in "Auditory features and affinities of the Eocene bats *Icaronycteris* and *Palaeochiropteryx* (Microchiroptera, incertae sedis)" for a 1987 edition of *American Museum Novitates*.

55. The tiger moth, *Bertholdia trigona,* is the only animal in the world known to have this ability, according to «How do tiger moths jam bat sonar?» from the *Journal of Experimental Biology* in 2011.

56. The barbastelle bat, *Barbastella barbastellus*, makes calls up to 100 times lower in amplitude than other bats that catch prey while in flight, researchers reported in "An aerial-hawking bat uses stealth echolocation to counter moth hearing" for *Current Biology* in 2010.

57. Beale wrote this in 1839 in admonishment of another writer, Abbe Lecoz, who reported that the sperm whale emitted "terrible groans when in distress."

58. Hal Whitehead's *Sperm Whales: Social Evolution in the Ocean* paints a beautiful and detailed picture of "the great leviathans."

59. That was based on just fourteen hours of underwater sound recordings, published as "The monopulsed nature of sperm whale clicks," for the *Journal of the Acoustical Society of America* in 2003. A larger sample might produce even louder sounds.

60. Eric Wagner covered this masterfully for *Smithsonian* magazine in 2011 in "The Sperm whale's deadly call."

61. Fifty-seven percent of the creaks were produced in the deepest 15 percent of the whales' dives, the researchers reported in "Sperm whale behavior indicates the use of echolocation click buzzes 'creaks' prey capture" for *Proceedings of the Royal Society B: Biological Sciences* in 2004.

62. The basic idea is that the features that make something like sea grass visible to sonar may be quite different from the features that make a fish, a sand ripple, or a sunken piece of a ship visible, according to "Environmentally-adaptive target recognition for SAS imagery," presented at the 2017 SPIE Defense + Security Conference.

63. The researchers believe that these young whales hadn't had a chance to learn to adopt alternative navigational strategies when they got lost, according to "Solar storms may trigger sperm whale strandings: explanation approaches for multiple strandings in the North Sea in 2016" in the *International Journal of Astrobiology* in 2017.

64. Today the global population is probably in the hundreds of thousands, reports the IUCN.

CHAPTER VI: THE TOUGH GET GOING

1. There were other dinosaurs that looked a lot more like the raptors from *Jurassic World.*

They just didn't happen to be velociraptors, as Brusatte reported in "New fossil reveals velociraptor sported feathers" for *Scientific American* in 2015.

2. In a fight between Chuck Norris and a water bear, I do profess, the latter would win with seven hands tied behind its back.

3. We might have as little as 100 years left here, Hawking has predicted.

4. When *Washington Post* reporter Ben Guarino wrote this up in 2017, the headline said it all: "These animals can survive until the end of the Earth, astrophysicists say."

5. "Tardigrades are as close to indestructible as it gets on Earth," Batista told *The Harvard Gazette* in 2017.

6. That will likely happen some time in the next 3.5 billion years.

7. Leaving them "at risk of population reductions or even extinction," according to "Will the Antarctic tardigrade *Acutuncus antarcticus* be able to withstand environmental stresses due to global climate change?" for the *Journal of Experimental Biology* in 2018.

8. Want to transfer life from one planet to another? Some NASA researchers believe the best bet isn't going to be humans, but tardigrades, an option discussed in "The tardigrade *Ramazzottius varieornatus* as a model animal for astrobiological studies" for *Biological Studies in Space* in 2008.

9. "These findings indicate the relevance of tardigrade-unique proteins to tolerability and tardigrades could be a bountiful source of new protection genes and mechanisms," Kunieda and his team wrote in "Extremotolerant tardigrade genome and improved radiotolerance of human cultured cells by tardigrade-unique protein" for *Nature Communications*, in what may have been the scientific understatement of 2016.

10. I was shocked and honored to win a Kavli Award for science journalism alongside Shubin in 2014. My award came for a story I wrote about Pando, the world's largest clonal organism, while Shubin's came for the three-part PBS series that was based on his book. I've never been so humbled.

11. Cautiously, the researchers who sequenced the coelacanth genome wrote in "Elephant shark genome provides unique insights into gnathostome evolution," for *Nature* in 2014, that their analyses indicate evolution in the coelacanth lineage "has occurred at a relatively slow rate, similar to that of non-mammalian tetrapods."

12. I've told Quong that if he ever gets another chance to swim with the elephant sharks, and needs someone to hold his gear, I'll be on the first plane to Australia. His remarkable video of the elephant shark release was published on YouTube in 2013.

13. It's important to note, though, that fifteen years after announcing the completion of a project to sequence the first human genome, scientists were still working on filling in some rather pesky gaps in the code. Sharon Begley covered this in 2017 for *STAT News* in the article "Psst, the human genome was never completely sequenced. Some scientists say it should be."

14. The North American beaver, *Castor canadensis*, had its genome sequenced in honor of the 150th anniversary of the founding of Canada—an important occasion, to be sure, but not one that carries any sort of scientific weight.

15. In "Conservation of all three p53 family members and Mdm2 and Mdm4 in the cartilaginous fish," the researchers wrote that their analysis showed that the elephant shark genome encodes three members of the p53 gene family: p53, p63, and p73. The paper was published in *Cell Cycle* in 2011.

16. Further exploration of the ways in which genes have evolved in species like the elephant shark and the naked mole rat, which is long-lived and resistant to cancer and many other diseases, "could bring novel insights into treating human medical issues," Michael Furlong and Jae Young Seong wrote in the journal *Biomolecules & Therapeutics* in 2017.

17. "Streak the Wonder Fish" learned to swim through hoops, eat from my fingers, and follow a pencil in a figure eight. Streak is buried next to a tree in the backyard of my boyhood home.

18. Hugo's book, *Strange Biology: Anomalous Animals, Mutants, and Mad Science*, is required reading for anyone who loves weird science.

19. "A creature that can regenerate lost body parts is strangely reminiscent of a *Spider-Man* backstory,» Hugo wrote for *Newsweek* in 2018.

20. In "Mexican salamander helps uncover mysteries of stem cells and evolution," Phys.org reported in 2010 that University of Nottingham researcher Andrew Johnson explained his preference for studying axolotls because, unlike many other frogs, fish, flies, and worms that are used in the lab, they have pluripotent cells. Like humans, their embryonic stem cells can become any other kind of cell.

21. "They are also distributed so commonly to labs for research that they are basically the white mice of amphibians," Sam Schipani wrote for *Smithsonian* magazine in 2018.

22. Writing of "A Tale of Two Axolotls" in *BioScience*, three axolotl researchers declared "a concerted global effort is needed to protect and manage this irreplaceable species in natural and laboratory environments." That was in 2015. At a global level, little has changed in the years since.

23. There are as many ways to measure slowness as there are to measure speed. In this case, we're talking about the mammal that is slowest, in absolute terms, when moving at its fastest speed.

24. "People say, what's in a name? Well, quite a lot when it is a synonym for a deadly sin," zoologist Lucy Cooke wrote in *The Day*. "Yes, the sloth was damned from the moment it was named after one of the world's most wicked transgressions." Cooke deserves great credit for having extricated from the dusty shelves of biological history the mean-spirited quote from Georges-Louis Buffon, "or buffoon as I like to think of him," she wrote.

25. They weren't hard to find, Pauli told me. "They don't go far."

26. There seems little way that sloths could survive with fewer calories. They live life on the edge.

27. Sex, as it turns out, is just about the only thing sloths do with any speed whatsoever. An entire sexual encounter might last as little as five seconds. "By all intents and purposes," Iva Roze Skoch reported for PRI's *GlobalPost* in 2012, "sloths might just take the honors for being every female species' worst nightmare."

28. I'm pretty sure that's what she said, at least. This, my dear Cambodian American friend Chhun Sun tells me, means "you don't understand."

29. In March of 2018, University of Iowa researcher Andrew Forbes and his colleagues made a compelling argument that wasps, not beetles, are a more diverse group of animals in "Quantifying the unquantifiable: Why Hymenoptera—not Coleoptera—is the most speciose animal order."

30. For a long time, scientists believed that the smallest insect was even smaller, per a description of a species of beetle called *Ptilium fungi* from John Eatton Le Conte in 1863. Le Conte's estimate was far more general than the way it was later cited, though. He had written that *P. fungi* was "scarcely more than 1-100 of an inch," but later measurements showed this to be incorrect. A more precise record was established in "How small is the smallest? New record and remeasuring of Scydosella musawasensis Hall, 1999 (Coleoptera, Ptiliidae), the smallest known free-living insect," which was published by *ZooKeys* in 2015.

31. The study "Ninety-eight new species of *Trigonopterus* weevils from Sundaland and the Lesser Sunda Islands," published by *ZooKeys* in 2014, included ninety-nine photos of weevil penises, differences in which helped the researchers identify many of the species as unique.

32. One clade in particular, Polyphaga, appears to have had a family-level extinction rate of zero for most of its evolutionary history, according to "The fossil record and macroevolutionary history of the beetles," published in *Proceedings of the Royal Society B, Biological Sciences* in 2015.

33. At any time, it is estimated that there are some 10 quintillion individual insects alive, according to *Smithsonian*. In the United States alone there are more than 90,000 known species.

34. Wallace was brutal. "What the Maine Lobster Festival really is," he wrote, "is a midlevel county fair with a culinary hook."

35. The first sushi restaurant didn't come to the United States until 1966, and it was almost exclusively frequented by Japanese immigrants. So how did sushi become such an American success story? The answer, according to Trevor Corson's *The Story of Sushi: An Unlikely Saga of Raw Fish and Rice*, includes the TV miniseries *Shōgun*, emerging ideas of healthy foods, and the Californication of foreign cuisine.

36. Martin's book is a delightful journey into "the next big trend in the global food movement."

37. In 2009, Rachael Rettner wrote about Gillooly for *LiveScience* in "Insect colonies function like superorganisms."

38. In chart after chart, the various colonies hugged the relational lines created by unitary organisms in "Energetic basis of colonial living in social insects," published in the *Proceedings of the National Academy of Sciences* in 2010.

39. Collectives react in ways that are strikingly similar to the nervous systems of individual organisms, according to "Differentiated anti-predation responses in a superorganism" from *PLOS One* in 2015.

40. Contrary to popular belief, the queen doesn't give orders, Tovey wrote for *The Conversation* in 2017. Each ant controls itself, in mind-bogglingly complex coordination with other ants.

41. Great apes, including humans, are some of the few exceptions. Giraffes might be exceptions to this rule, too, but no one seems to know for sure.

42. Flannery's review of Bert Hölldobler and Edward Wilson's 2009 treatise on ants, *The Superorganism*, in the *New York Review of Books*, was as good as the book itself. Flannery called *The Superorganism* "a profoundly important book with immediate relevance for anyone interested in the trends now shaping our own societies." He's right.

CHAPTER VII: DEADLY SERIOUS

1. I thought at first this might have been a story my guide was telling me to keep me in line. Later, Dirk Donath's disappearance was confirmed in the media.

2. Want to know what's actually most likely to kill you? You can find that and more in "Global and regional mortality from 235 causes of death for 20 age groups in 1990 and 2010: a systematic analysis for the Global Burden of Disease Study 2010" from the *Lancet* in 2012.

3. Spiders are just one of many animals we fear far more than the numbers indicate we should. CNN reporter Jacqueline Howard, in response to an alligator attack that claimed the life of a boy in Florida in 2016, noted the death was "without a doubt horrific—and extremely rare."

4. If you believe in a god who smites people, consider this: Lightning kills more than three times more men than women, according to the National Oceanic and Atmospheric Administration.

5. The Insurance Information Institute's handy mortality risk chart is fun reading for anyone who drives, flies, swims, or lives in a country in which guns outnumber people.

6. The tourism industry hailed the government's decisions, but environmentalists did not, as Reuters reported in "Spate of Australia shark attacks could take a bite out of tourism" in 2017.

7. The United States Department of Agriculture's handbook *Plants Poisonous to Livestock in the Western States* reads like a textbook from one of Professor Pomona Sprout's classes.

8. This kind of death was described in "Pyrrolizidine alkaloidosis in a two month old foal" in the *Journal of Veterinary Medicine* in 1993.

9. These benefits include the development of animal models, isolation of novel compounds, and insights into the biological and molecular mechanisms of plant-based chemicals, according to "The good and the bad of poisonous plants: an introduction to the USDA-ARS Poisonous Plant Research Laboratory," from the *Journal of Medical Toxicology* in 2012.

10. Matthew Herper wrote about "The curious case of the one-eyed sheep" for *Forbes* in 2005.

11. "Many plants evolve toxins to ward off bugs and animals," Brian Maffly wrote for the *Salt Lake Tribune* in 2017, "and biomedical scientists have discovered these can be harnessed for medicinal uses," but first they have to find places where these plants can successfully grow.

12. According to "Anticancer potential of *Conium maculatum* extract against cancer cells in vitro: Drug-DNA interaction and its ability to induce apoptosis through ROS generation," published in *Pharmacognosy Magazine* in 2014, the researchers deduced that hemlock hinders the process of cell proliferation and cell cycle.

13. John Mann from Queen's University Belfast wrote of the history and potential of deadly nightshade in the 2008 article "Belladonna, broomsticks and brain chemistry" for *Education in Chemistry.*

14. Death generally takes place within 36 to 72 hours of exposure, according to the US Centers for Disease Control.

15. Phytoremediation may also be good for biodiesel production, medicinal product development, and carbon sequestration, according to "*Ricinus communis*: A robust plant for bio-energy and phytoremediation of toxic metals from contaminated soil," published by *Ecological Engineering* in 2015.

16. In one study, "Evaluation of *Ricinus communis L.* for the phytoremediation of polluted soil with organochlorine pesticides," published in 2015 by *BioMed Research International*, the plant showed a remediation effect of up to 70 percent.

17. There are scores of *Sinularia* species, also known as "leather corals," in the ocean. Some have been known for more than a century, but few have been examined by scientists, as detailed by "A new norcembranoid dimer from the Red Sea soft coral Sinularia gardineri," in the *Journal of Natural Products* in 1996.

18. "Why would a plant produce two types of substances which may act as physiological antagonists of each other?" Oné R. Pagán, the authors of *Strange Survivors*, asked in 2012. "This is kind of weird when you think about it; like giving a poison and its antidote at the same time." Later, we'll discuss how a plant called thale cress uses two conflicting chemicals.

19. If there's a guidebook for starting a tobacco business, it's *Tobacco: Production, Chemistry, and Technology*, which covers everything from seed to store and from storage to sales.

20. As of 2010, David Heath noted in the *Atlantic* in 2016, tobacco companies were still routinely arguing "that the nation's top-selling cigarette, once known as Marlboro Lights and now called Marlboro Gold, reduces the risk of cancer."

21. In 2017, finally, tobacco companies admitted they purposefully made cigarettes more addictive. Seriously, it wasn't until twenty-frickin-seventeen. That November, NPR's Alison Kodjak explained why for *All Things Considered.*

22. Jennifer Maloney of the *Wall Street Journal* explained how this worked to NPR's *All Things Considered* in 2017.

23. "It's a great time to be a cigarette company again," Maloney wrote in the *Wall Street Journal* in 2017. "Far fewer Americans are smoking, and yet U.S. tobacco revenue is soaring."

24. It was long thought that Euro-American traders were responsible for the spread of tobacco across the United States in the late 1700s. While that might be true when it comes to *N. tabacum,* archeologist Shannon Tushingham and her colleagues have demonstrated through the molecular analysis of ancient pipes that *N.*

quadrivalvis and *N. attenuate* were being used in what is now considered the American West at least 1,200 years ago. The team published these findings in the *Proceedings of the National Academies of Sciences of the United States of America* in 2018.

25. Counterintuitive findings like this usually do quite well in the media; for whatever reason this one flew under most science journalists' radars. That might have had something to do with the title of the study, which was published in 2017 in the journal *Bioorganic & Medicinal Chemistry*: "The tobacco cembranoid (1S,2E,4S,7E,11E)-2,7,11-cembratriene-4,6-diol as a novel angiogenesis inhibitory lead for the control of breast malignancies."

26. Writing for *Forbes* in 2016, Sally Satel, a psychiatrist specializing in addiction medicine who studies the intersection of medicine and culture, called these policies "absurd."

27. Their 2014 study, "Phosphoinositide-mediated oligomerization of a defensin induces cell lysis," was published in *eLife*, which at the time was a two-year-old open-access journal with a low impact factor relative to heavy hitters like *Nature*, *Science*, and *Cell*. A lot of research demonstrating beneficial uses for tobacco is published in journals like this.

28. "I am beginning to feel kindly towards the genus *Nicotiana,*" she wrote in 2014.

29. Please don't do this. Ever. But do feel free to listen to Tarvin describe the experience, which she did for NPR's *All Things Considered* in 2016.

30. Tarvin and her team, from the University of Texas, St. John's, and Harvard in the United States, and the University of Konstan in Germany, published their findings in "Interacting amino acid replacements allow poison frogs to evolve epibatidine resistance" for *Science* in 2017.

31. Sometimes, the best scientists can do is to "log" a species and take a few specimens, Carrie Arnold reported for *National Geographic* in 2017.

32. There are nearly as many species of amphibians categorized as threatened as those of birds and mammals together, a group of biologists warned in the journal *Sapiens* in 2012 in the article "The Amphibian Extinction Crisis—what will it take to put the action into the Amphibian Conservation Action Plan?"

33. The question now is whether these species face an even greater risk of human-driven extinction, according to the 2015 report "Antipredator defenses predict diversification rates" for the *Proceedings of the National Academy of Sciences.*

34. The results suggest "that how a species defends itself might be part of the puzzle of working out which species are in need of conservation efforts," Kevin Arbuckle of Swansea University told the *Express* in 2016.

35. The work, headed up by Stanford University chemist Justin Du Bois, may pave the way for subsequent studies aimed at understanding the many small molecules found in the toxins of frogs and other animals, according to 2016's "Asymmetric synthesis of batrachotoxin: enantiomeric toxins show functional divergence against NaV" in *Science.*

36. The structure of the toxin—and a partial synthesis—was first described in the *Journal of the American Chemical Society* in 1969.

37. With the price of one of the most effective rattlesnake antivenom treatments at $14,000 or more, you might rather just take your chances. (But please don't. Get the treatment and then "go to the mattresses" with the hospital and insurance companies.)

38. Like the sixty-year-old Kentucky man who was killed while handling a snake during a Sunday service at a Pentecostal church in 2015.

39. Also in 2015, a Missouri man who was twice bitten by a water snake died the following day after refusing to go to the hospital.

40. Another Missouri man died after voluntarily picking up a wild snake. Unsurprisingly, he was bitten.

41. An eighteen-year-old Texas man named Grant Thompson "died surrounded by animals that intrigued and fascinated him," according to his obituary in the *Austin American-Statesman* in 2015. A police investigation concluded he'd purposefully allowed his pet cobra to bite him multiple times while sitting in a car in the parking lot of a home improvement store in Austin, Texas.

42. Snakes just beat out spiders as the scariest animal in a 2014 survey by the British polling firm YouGov. Interestingly, the older a respondent was, the more likely they were to report being "very afraid" of snakes, while the number of those reporting tremendous fear of spiders decreased with age.

43. That work built on earlier research showing that lab-raised monkeys with no prior exposure to snakes had no fear of the reptiles, even though they could detect snakes more quickly than other harmless creatures. Lee Dye offered a good overview of this research for ABC News in 2011 in "Afraid of snakes? Scientists explain why."

44. The study was led by Stephanie Hoehl of the Max Planck Institute for Human Cognitive and Brain Sciences, Leipzig, Germany, and published with the I-am-sure-to-read-this title "Itsy bitsy spider . . . : infants react with increased arousal to spiders and snakes," in *Frontiers in Psychology* in 2017.

45. An estimated 5.4 million people are bitten by venomous snakes each year, according to the World Health Organization.

46. Isbell's book, *The Fruit, the Tree, and the Serpent*, suggests snakes gave humans an evolutionary nudge that made us what we are today.

47. In 2013, Isbell told NPR's *All Things Considered* how she began to wonder about the snake–human connection. The story involves a cobra, of course.

48. "Pulvinar neurons reveal neurobiological evidence of past selection for rapid detection of snakes" was published in the *Proceedings of the National Academy of Sciences* in 2013.

49. Interestingly, female and juvenile monkeys detected the snakeskin better than adult males.

50. At just four months old, the monkeys were able to recognize snakes and engage in anti-predator behavior, according to "Scales drive detection, attention, and memory of snakes in wild vervet monkeys (*Chlorocebus pygerythrus*)," from the journal *Primates* in 2017.

51. Test subjects showed the largest early posterior negativity reactions to snakes. Spiders triggered medium reactions. Slugs barely moved the needle. "Testing the snake-detection hypothesis: larger early posterior negativity in humans to pictures of snakes than to pictures of other reptiles, spiders and slugs," from *Frontiers in Human Neuroscience* in 2014, was eye-opening reading.

52. To me, this is one of the most interesting findings. Some humans have developed fear, but all humans, it seems, carry attentiveness to snakes in our psyches, as exhibited in "Snake scales, partial exposure, and the Snake Detection Theory: a human event-related potentials study" in *Science Reports* in 2017.

53. "A global biodiversity crisis is threatening the very snake populations on which hopes for new venom-derived medications depend," the authors wrote in "Snake venom: from fieldwork to the clinic" for *Bioessays* in 2011.

54. "The harsh fact is that the continent is largely devoid of safe, effective, and affordable treatments for something that is eminently treatable," Williams wrote in "Snake bite: a global failure to act costs thousands of lives each year" for the *British Medical Journal* in 2015.

55. They heard back from 43 percent of the health agencies, 27 percent of the poison centers, and 13 percent of the manufacturers. That's dreadful. The conclusions in "Needs and availability of snake antivenoms: relevance and application of international guidelines" in the *International Journal of Health Policy and Management* are thus couched in caveats.

56. The project aimed to create "a strong impact for European competitiveness, at both an academic and economic level," according to the Venomics report summary from the Community Research and Development Information Service of the European Union. Saving lives was secondary.

57. Jenny Bryan covered the history of this drug in "From snake venom to ACE inhibitor—the discovery and rise of captopril" for *Pharmaceutical Journal* in 2009.

58. Later some of the indigenous groups accused researchers of "biopiracy," and Brazil began to crack down on companies that patent products without compensating the nation or its indigenous peoples, according to a 2010 Reuters article titled "Brazil to step up crackdown on 'biopiracy' in 2011."

59. Because Nobels are often given for work that is many years old, laureates' speeches are often focused not on what they are being honored for, but on what they are doing now.

60. Thus we now have very large libraries of toxin variants that can be tested against myriad diseases, thanks in part to a scientist named Zoltan Takacs, who founded the World Toxin Bank. I learned about the toxin bank from Kath Nightingale's "The bite that cures: how we're turning venom into medicine" for the BBC's *Science Focus* in 2016.

61. Data journalist Sascha Steinhoff has created a very useful and searchable database of venomous snakes at http://snakedatabase.org.

62. Researchers from Monash University in Australia reported their findings on the inland taipan in "Some pharmacological studies of venom from the inland taipan (*Oxyuranus microlepidotus*)" for the journal *Toxicon* in 1998.

63. The researchers, from the Institute of Molecular and Cellular Pharmacology in France, reported their findings to *Nature* in 2012 in the article "Black mamba venom peptides target acid-sensing ion channels to abolish pain."

64. It also has an elegant chemical structure, viewable on the DrugBank databank, which is supported by the Canadian Institutes of Health Research, among other organizations.

65. The researchers, from Tezpur University, published their findings in "Proteomics and antivenomics of Echis carinatus carinatus venom: Correlation with pharmacological properties and pathophysiology of envenomation" for *Scientific Reports* in 2017.

66. The World Health Organization estimates that there are 3,000 species of snakes in the world. Only about 600 are venomous, and only about a third of those are suspected to be "medically important."

67. Death certificates aren't required in many countries where box jellyfish attacks are common, Stuart Fox wrote in "How deadly Is the box jellyfish?" for *LiveScience* in 2010.

68. "Jellyfish and other cnidarians are the oldest living venomous creatures," Fry told Catherine Paddock of *Medical News Today* in 2015, "but research has been hampered by a lack of readily obtainable venom harvested in a reproducible manner."

69. It takes half an hour in the magic spinner, according to Fry's technique, which he described in "Firing the sting: chemically induced discharge of cnidae reveals novel proteins and peptides from box jellyfish *(Chironex fleckeri)* venom" for the journal *Toxins* in 2015.

70. In "Stonefish toxin defines an ancient branch of the perforin-like superfamily," the researchers, from Australia's Monash University, gave the world its first high-resolution insights into how the Membrane Attack Complex-Perforin/Cholesterol-Dependent Cytolysin domains interact in the *Proceedings of the National Academy of Sciences* in 2015.

71. "The largest and most clinically important pharmacopoeia of any genus in nature," Center for Health and the Global Environment founder Eric Chivian told *National Geographic* in 2003.

72. The *Washington Post's* Sarah Kaplan described this scene beautifully in an August 13, 2017, article about Marí's work.

73. C. Renée James told this story and more like it in her delightful book *Science Unshackled: How Obscure, Abstract, Seemingly Useless Scientific Research Turned Out to Be the Basis for Modern Life.*

74. Daisy Yuhas noted in *Scientific American* in 2013 that the cone snail's toxins could be used to treat epilepsy and depression, too.

75. "Spiders exist in the most northern islands of the Arctic, the hottest and most arid of deserts, at the highest altitudes of any living organisms, in the depths of caves, in the intertidal zone of ocean shores, in bogs and ponds, on high, arid moorlands, sand dunes, and floodplains," the Canadian arachnologist wrote in "Ecology of the true spiders (Araneomorphae)" for the *Annual Review of Entomology* in 1973.

76. That's about 1 percent of the net production of meat for the entire terrestrial Earth, according to "An estimated 400-800 million tons of prey are annually killed by the global spider community," published by researchers from Switzerland, Sweden, and Germany in *The Science of Nature* in 2017.

77. Ingraham's reports on "all things data" for the *Washington Post* are always fascinating and often hilarious.

78. Okay, the world would be a nicer place with a few more butterflies. But still.

79. However, "if the researchers are to develop a drug or therapy involving compounds from spider venom," Lecia Bushak wrote for *Medical Daily* in 2015, "it will likely be quite some time before they can be tested in a clinical setting."

80. Fun fact from Rachel Nuwer's "Could spider venom be the next Viagra?" on Smithsonian.com in 2012: A persistent, intensely painful erection is known as priapism.

81. Former US vice president and Nobel Prize winner Al Gore loves talking about this, but he doesn't explain it as well as Geoffrey Fattah did for the *Deseret News* in 2011 in the story "A web of possibilities: Utah researcher uses goats to make one of the strongest known substances."

82. Gates was passionate and quite funny in his eighteen-minute talk, which is worthwhile watching for anyone who wants an optimistic primer on the battle to end malaria.

83. I've never seen, or heard, a better example of the ways in which privilege impacts perspective.

84. As of 2013, the World Health Organization put the spending on malaria control at $1.8 billion; on hair loss, at $2 billion, according to "Medical research: the bald truth," a 2013 editorial in *The Guardian.*

85. "Antimalarial drug resistance," a good primer on antimalarial drug resistance, was written by Nicholas White of the Mahidol University in Thailand in 2004 for *The Journal of Clinical Investigation*.

86. Jerry Alder's article "Kill all the mosquitoes?!" noted that gene-editing technology had advanced to the point that scientists had the ability to wipe out all of the known carriers of malaria and Zika. The unresolved question, he noted, was, "Should they?"

87. Fang asked questions in "A world without mosquitoes" that a lot of scientists hadn't even thought about yet. It's still thought-provoking reading.

88. Almost immediately, other countries lined up to volunteer to be Oxitec's next guinea pigs, according to "GM mosquitoes wipe out dengue fever in trial," from *Nature*'s news blog in 2010.

89. Oxitec officials rejected the notion that their trial was secretive. It was well known, they said, on the island, "but just not picked up internationally," Martin Enserink wrote for *Science* in 2010.

90. And, in an example of Darwinism in action, there is some evidence that female mosquitoes have figured out how to avoid the genetically engineered males, Alder reported.

91. Framing is everything. The authors of "Highly efficient Cas9-mediated gene drive for population modification of the malaria vector mosquito *Anopheles stephensi*" began their 2015 study report in the *Proceedings of the National Academy of Sciences* not by introducing the background of the technology they were using, but the problem they were battling: a disease that kills hundreds of thousands of people each year.

92. "Many people think it'll be efficient and predictable. But that's not the case here. We need to know how to talk about to the public, so they understand the risks," she told journalist Alex Zielinski from *ThinkProgress* in 2016.

93. Pugh, though, seemed on the fence about the question. "Humans have been selectively breeding both plants and animals for hundreds of years," he wrote in *The Conversation* in 2015, "and this can be viewed as an indirect form of genetic modification that we do not find morally problematic."

94. *The Worldwide Threat Assessment of the US Intelligence Community* might put you to sleep. Once asleep, though, it might also give you nightmares.
95. In 2017, Cornell University's Robert Reed once told Ed Yong that CRISPR had turned "the biggest challenge of my career . . . into an undergraduate project."
96. Yes, you can get a fairly sophisticated CRISPR lab sent to your home; I've got one. For now the chief argument suggesting that people won't use these to create malicious transgenic organisms is that it's actually easier to create other sorts of bioweapons that could do even more harm. That's not reassuring.
97. UC Berkeley professor John Marshall's primer on the agreement and its relationship to bio-engineered bugs, "The Cartagena protocol and releases of transgenic mosquitoes," is important reading.

CHAPTER VIII: SMARTER ALL THE TIME

1. The entire interaction, video of which was included as supplemental information with the 2018 article "Precocious development of self-awareness in dolphins" in *PLOS One*, lasts just about thirty seconds. When I watch it, I don't see much more than a truly adorable dolphin. But I'm not a dolphin expert.
2. "It's easy to prove that dolphins are capable of doing something—it only takes one dolphin to do that," research analyst Hannah Salomons told me. "But showing dolphins *can't* do something is a lot harder. We had to run this experiment with a lot of dolphins. There are three buckets, and they guessed right about a third of the time."
3. "Measurement of hydrodynamic force generation by swimming dolphins using bubble DPIV" was published in the *Journal of Experimental Biology* in 2014.
4. And sharks, of course, attack and kill a lot of dolphins, too, as Joshua Rapp Learn wrote in "Tracking the scars of dolphin-shark battles" for *Hakai Magazine* in 2017.
5. "The dolphins in this photo have been backed into a corner, with fishermen and certain death on one side, and the net separating them from freedom on the other," Bridgeman wrote for *Ecologist* in 2013.
6. We've known this for quite some time—it was first reported in "The limbic lobe of the dolphin brain: a quantitative cytoarchitectonic study" for *Journal für Hirnforschung* in 1982.
7. His book *In Defense of Dolphins: The New Moral Frontier* mixes philosophy and science in thought-provoking ways.
8. It's tempting to think of this as "values-driven behavior." If you do, you need to take the good with the bad. Typical circumstances for dolphins also include actions that we would call infanticide and rape.
9. This remarkable recall, James Ritchie wrote for *Scientific American* in 2009, may be a factor in the elephant's survival—another anchor keeping them on the top side of Cope's Cliff.
10. In "Post-traumatic stress? It doesn't even exist!" journalist Paul Strudwick reported

Bhugra's claims that "the condition is not a true mental illness but instead is being diagnosed as a result of the influence of 'insurance firms and drug manufacturers.'"

11. The national studbooks for both African and Asian elephants show the majority of elephants on display throughout the United States were captured in the wild as juveniles.

12. In most cases, Schobert said, the calves he purchased for his zoos were still drinking their mother's milk when they were separated from their herds, and had to be trained to use a bottle. I first reported this for the *Salt Lake Tribune* in 2008.

13. "Parental olfactory experience influences behavior and neural structure in subsequent generations," by Emory University researchers Brian Dias and Kerry Ressler, was published in *Nature Neuroscience* in 2013.

14. "The study of transgenerational epigenetic inheritance by definition requires the maintenance and breeding of more than one generation of organisms and the significant costs in time, money, space, *etc.*, that are required," he wrote in "Epigenetic inheritance and its role in evolutionary biology: re-evaluation and new perspectives" for *Biology* in 2016.

15. Not "octopi," since "octopus" does not come from Latin. And even though the word is Greek in origin, most folks don't say "octopodes," either. (But if you *want* to say it that way, the pronunciation is ock-TOP-uh-deez.)

16. Godfrey-Smith first wrote about this in a 2016 essay for the *New York Times* called "Octopuses and the Puzzle of Aging." In it, he speculated that if octopuses could evolve to tilt the odds in their favor a little more, their lifespans might increase, although maybe not to as long as a human lifespan. "And when one contemplates the thought of a century-old octopus," he wrote, "perhaps that's just as well."

17. Where does all that intelligence come from? The octopus genome is slightly smaller than the human genome, but it has 50 percent more genes, and has hundreds of novel genes not found in any other animal, *Independent* science editor Steve Connor reported in "Armed with 10,000 more genes than humans" in 2015.

18. In a 1998 article for *Current Biology*, Brenner told fellow biologist Lewis Wolpert that he was envious of Charles Darwin, "but it is impossible to begrudge him his success and demand that he should have waited a century or so to allow me a fair chance to compete with him."

19. In "The octopus genome and the evolution of cephalopod neural and morphological novelties," published in *Nature* in 2015, Brenner's team reported on their successful sequencing of the genome of the California two-spot octopus, *Octopus bimaculoides*.

20. In "Octopus inspires AI robots on a mission," published by *Seeker* in 2016, science journalist Tracy Staedter wrote about a cephalopod-based AI project underway at Raytheon in Aurora, Colorado.

21. Females from the genus *Neotrogla* have an elaborate structure called a "gynosome," which fits into the male genital chamber. Copulation takes up to seventy hours. "Nothing similar is known among sex-role reversed animals," the authors wrote in "Female penis, male vagina, and their correlated evolution in a cave insect" for *Current Biology* in 2014.

22. *Physarum polycephalum* is a large amoeboid that changes its shape as it crawls. If food is placed at two different points in a maze, the researchers showed, *P. polycephalum* will find the minimum-length distance between the food, according to «Intelligence: maze-solving by an amoeboid organism» for *Nature* in 2000.

23. Fascinatingly, the amoebas do eventually "forget" their last turn—after about ten minutes of inactivity, according to "Persistent cell motion in the absence of external signals: a search strategy for eukaryotic cells" in *PLOS One* in 2008.

24. We know that circadian rhythms are powerful forces in all sorts of organisms. They are as old as the connection between life on Earth and our sun. But in "Amoebae anticipate periodic events" for *Physical Review Letters* in 2008, the research team demonstrated an anticipated response to rhythm in single-celled organisms after just a few hours of "training."

25. First conceived in 1971 by the influential electrical engineer Leon Chua, memristors remained a theoretical construct until 2008, when a team from Hewlett-Packard demonstrated that memristance occurs in natural nanosystems.

26. *P. polycephalum* probably isn't particularly special. "These biological memory features are likely to occur in other unicellular as well as multicellular organisms, albeit in different forms," the authors wrote in "Memristive model of amoeba's learning" in *Physical Review E.* in 2010.

27. Jacek Krywko did an excellent job at breaking this down in "Electronic synapses that can learn signal the coming of the first real artificial brain" for *Quartz* in 2017.

28. "With that much power, it must be doing some important work. Curing cancer, perhaps?" Andy Boxall wrote for *Digital Trends* in 2012. "Sadly not, it's going to be running simulations of the effectiveness of nuclear weapons, and working out how to safely extend their life." This is the best (and worst) example of humans exploiting their best minds to support their worst impulses.

29. When Rebecca Boyle pointed this out in "Simulated brain ramps up to include 100 trillion synapses" for *Popular Science* in 2012, I felt proud to be human for a few moments.

30. Their paper, "Bitcoin emissions alone could push global warming above 2°C," published in *Nature Climate Change* in 2018, is terrifying reading.

31. Grollier's talk, called "Realizing a brain on a chip," will blow your mind.

32. This opens paths to "unsupervised machine learning," the authors hypothesized in "Learning through ferroelectric domain dynamics in solid-state synapses" for *Nature Communications* in 2017. What happens after that is anyone's guess, and could either give us the help we need to survive, with AI increasingly responsible for the second-by-second decision making needed to keep our world running, or put the final nail in our coffin, as superintelligent machines realize the world would be a much better place without us.

33. And yet they exhibit behaviors that are often as complex, and sometimes far more complex, than those of much larger species, the authors pointed out in "The allometry of brain miniaturization in ants," published in *Brain Behavior and Evolution* in 2011.

34. A great analogy, served up even better by Bryan Walsh in "Your ant farm is smarter than Google" for *Time* in 2014.

35. John Koetsier explained how this works in "How Google searches 30 trillion web pages, 100 billion times a month" for *Venture Beat* in 2013.

36. In 2014, Kurths told the *Independent* that the mathematical model used in his study of ant search algorithms might also apply to other homing animals, like albatrosses, the world's largest seabird.

37. In "Learning universal computations with spikes" published in *PLOS Computational Biology* in 2016, researchers from the Netherlands and the United States showed how short electrical impulses in animal brains help generate a picture of the world that is both in chaos and broadly predictable.

38. In "A modified BFGS formula using a trust region model for nonsmooth convex minimizations" published in *PLOS One* in 2015, researchers proposed a new algorithm for identifying the best solution out of many possible solutions for certain nonlinear problems.

39. In "A hybrid optimization method for solving Bayesian inverse problems under uncertainty" published in *PLOS One* in 2015, a team of Chinese researchers offered a new method for history matching, the process of building multiple numerical models to account for measured data.

40. In "Army ants dynamically adjust living bridges in response to a cost–benefit trade-off" for *Proceedings of the National Academy of Science* in 2015, the authors suggested there could be "potential implications for human engineered self-assembling systems."

41. Oh, and they can't even see, Morgan Kelly from Princeton University explained in "Ants build 'living' bridges with their bodies, speak volumes about group intelligence" in 2015.

42. That's what he told Bryan Walsh at *Time*.

43. One exception is the Australian bulldog ant *Myrmecia nigriceps*, which has significantly better eyesight than other ants, according to "Attack behavior and distance perception in the Australian bulldog ant *Myrmecia nigriceps*," published in 1985 by the *Journal of Experimental Biology*.

44. "The ant odometer: stepping on stilts and stumps" was published in the journal *Science* in 2006.

45. Writing in the journal *Science* in 2016, the authors of "Prospective representation of navigational goals in the human hippocampus" concluded that a hippocampal-cortical network supports a "mind map" whenever we make a navigational goal and seek it out.

46. In "We've been looking at ant intelligence the wrong way," for *The Conversation*, Antoine Wystrach of the University of Sussex also argues that, although "it seems intuitive to start with our own assumptions about human intelligence," we can't answer questions about insect intelligence with a "top-down approach."

47. "The average velocity of the ants is almost independent of their density on the trail," the authors observed in "Trafficlike collective movement of ants on trails: absence of a jammed phase" for *Physical Review Letters* in 2009. That's amazing.

48. When self-driving cars arrive in a construction zone, Aarian Marshall wrote for *Wired* in 2017, they "can't even."

49. Sherman's song is based on the tune of Amilcare Ponchielli's "Dance of the Hours."

50. *The Secret Life of Plants* is actually a pretty good read. Take it with a grain of salt, though. Or maybe a whole mountain of salt.

51. *Rolling Stone* didn't agree with my assessment. The brilliant music critic Ken Tucker acknowledged it was "full of tiny pleasures" but also called it "uneven" and marred by "bloated tedium," adding "one person's nectar is another's Karo syrup." And whether it was the music or the subject matter, a lot of people really loved the movie. My amazing editor, Leah Wilson, is pretty sure she watched it in a biology class in the 1990s.

52. "The fact that the term 'neuron' is derived from a Greek word describing a 'vegetable fiber' is not a compelling argument to reclaim this term for plant biology," the authors wrote in "Plant neurobiology: no brain, no gain?" for *Trends in Plant Science* in 2007.

53. Michael Pollan wrote about the debate for *The New Yorker* in 2013. "Depending on whom you talk to in the plant sciences today," he wrote, "the field of plant neurobiology represents either a radical new paradigm in our understanding of life or a slide back down into the murky scientific waters last stirred up by *The Secret Life of Plants.*"

54. "This relatively long-lasting learned behavioral change," Gagliano and her collaborators wrote in the journal *Oecologia* in 2014, "matches the persistence of habituation effects observed in many animals."

55. Robert Krulwich (of *Radiolab* fame) crafted a masterfully understated lead in his piece about Gagliano's work for *National Geographic* in 2015. "There's this plant I've heard about that had a really bad afternoon a few years ago," he wrote.

56. Some scientists do believe we are getting closer to understanding how memory works in the human brain, Elizabeth Dougherty explained in "Map in your mind" for Boston University in 2016.

57. "Learning by association in plants," published in *Scientific Reports* in 2016, is beautiful for its simplicity. A talented middle school student could replicate this process with another species of plant for a science fair project. (Hey, Spike: That's a hint.)

58. A lot of human decision-making processes work in similar ways, and even using similar chemical signalers, the authors of "Temperature variability is integrated by a spatially embedded decision-making center to break dormancy in *Arabidopsis* seeds" wrote in the *Proceedings of the National Academy of Sciences* in 2017. Perhaps plants—eukaryotes just like us—really are "slow animals."

59. "The biomass distribution on Earth," published in the *Proceedings of the National Academy of Sciences* in 2018, offers plenty of humbling perspectives on our relative importance to this world. For example, every animal in the world combined accounts for less than a third of the biomass of the planet's archaebacteria.

60. Gibson writes eloquently on the intersections of art, philosophy, and science, including in "Pavlov's plants: new study shows plants can learn from experience" for *The Conversation* in 2016.

CONCLUSION: THE NEXT SUPERLATIVE DISCOVERY IS YOURS

1. Their collective screeches are among the loudest animal sounds ever recorded, according to Brad Erisman and Timothy Rowell in "A sound worth saving: acoustic characteristics of a massive fish spawning aggregation," published in *Biology Letters* in 2017.

2. Flesher's "Analysis of *Populus tremuloides* clonal variation and delineation in the Ottawa National Forest" is a beautiful piece of science.

INDEX

A

Abyssinian cat, 139
Acinonyx jubatus. see cheetahs
"Acoustic classification of Australian frogs based on enhanced features and machine learning algorithms," 4
acoustic systems, bio-inspired, 181
Actual Size (Jenkins), 293–294, 307*n*6
Acutuncus antarcticus (Antarctic tardigrade), 192
Adidas, 250–251
aerobic energy metabolism, in humans and elephants, 22
African bush elephant *(Loxodonta africana),* 18, 311*n*23
  common ground between humans and, 21–22
  preserving sperm of, 29
African clawed frog *(Xenopus laevis),* 6
African elephants, 16. *see also* African bush elephant *(Loxodonta africana)*
  Cope's Rule and, 17–18
  hybridization of Asian elephants and, 309*n*3
  preserving sperm of, 29
  rock hyrax and, 63
African forest elephant *(Loxodonta cyclotis),* 309*n*2
age. *see also* longevity; oldest
  of bowhead whales, 115
  of Pando, 95–100
  size and, 110
  of trees, determining, 96–97, 110–112
age-related illnesses, 6
aging, 113
  of African clawed frog, 6
  in animals and in humans, 6–7
  of bristlecone pines, 113–115
  comparative genomics of, 115
  of frogs, 6–7
  of *Hydra vulgaris,* 116–117
  negligible, 320*n*23
  telomeres' role in, 6
Alder, Jerry, 340*n*86
alfalfa, 250
allergies
  adults with, 316*n*28
  to penicillin, 8
Alouatta species (howler monkeys), 159–163
Alroy, John, 310*n*12
ambassadors, scientific, 9
Amboseli National Park, 269, 270
Ambystoma mexicanum (axolotls), 200–203, 332*n*20
Ameerega shihuemoy (species of poison dart frog), 233
American alligator, 173
American bison, 142

American cockroach *(Periplaneta americana),* 146, 323*n*27
American Gut study, 75–76
Ammonicera minortalis (water snail), 93–94
amoebas, 278
amphibians, 174. *see also specific amphibians*
endangered habitats of, 233
poisonous, 233–234
sounds made by, 174–175
species of, 336*n*32
Ancient Bristlecone Pine Forest, 108
anemones, 245, 246
Anger, Natalie, 323*n*25
anhydrobiosis, 194
animals. *see also specific animals*
deadly to humans, 221–222
gene-drive-altered, 254–256
killed and eaten by humans, 248
killed and eaten by spiders, 248–249
scientific reports on care and feeding of, 308*n*10
size and speed of, 136
in total biomass, 345*n*59
Annals of the Entomological Society of America, 179
Anopheles gambiae (mosquito), 253. *see also* mosquitoes
Antarctic tardigrade *(Acutuncus antarcticus),* 192
ant colonies, 214–217. *see also* ants
bridge building by, 282–283
collective intelligence of, 283
information processing in, 281–283
queen in, 333*n*40
as superorganisms, 215–217
antelopes, 142–143
anticoagulants, 244
Antilocapra americana (pronghorn), 141–145, 322*n*18
antimalarial drug resistance, 340*n*85
"Antioxidative responses of the tissues of two wild populations of Pelophylax kelpton esculentus frogs to heavy metal pollution," 4
antivenom manufacturers, 239–241
antivenom treatments, 336–337*n*37, 338*n*54
ants, 213. *see also* ant colonies; *specific types of ants*
brains of, 280
foraging behavior of, 281, 282
information processing by, 282, 284
intelligence of, 280–287
navigation by, 283–287
vision in, 283, 344*n*43
Anura, 7
apes, 20, 310*n*13, 334*n*41
apoptosis, 25
apple trees, 48
Arabidopsis thaliana (thale cress), 291
Arachnoscelis arachnoides (Colombian bush cricket), 170
Arbuckle, Kevin, 336*n*34
Archaea, 63
archosaurs, 172–173
Arctica islandica (quahog clam), 120–121
Armillaria ostoyae (humongous fungus), 297
army ant *(Eciton hamatum),* 283
Arnold, Carrie, 314*n*3, 336*n*31
artificial intelligence
cephalopod-based, 342*n*20
drone-based search and rescue, 276
human versus simulated brains, 279–280
single-celled organisms as key to, 277–278, 280
unsupervised machine learning, 343*n*32
artificial memory, 279
Arujo, Diego Gustavo Ahuanari, 231–232
Ashenafi, Zelealem Tefera, 139–140
Asian elephant *(Elephas maximus),* 16
hybridization of African elephants and, 309*n*3
preserving sperm of, 29
sperm of, 311*nn*21–23
stress in, 270
Asian macaques, 237
Asian tiger mosquito, 79, 81, 317*n*38
Asiatic cheetahs, 323*n*17
aspens. *see also* Pando (aspen colony)

bark of, 299–300
disruptors and, 104–106
formula for age of, 97
interconnectedness and, 103–108
investigating, 297–300
leaf differences among clones, 299
longevity factors for, 125–126
migration of, 52, 98–99
number of chromosome sets in, 100
oldest, 95–100
pollen fossils from, 99
quaking, 50–51, 54, 102, 125, 298
size of, 51–55, 297–298
species of, 298
in Yellowstone, 105
Association of Zoos & Aquariums, 27–28
associative learning, 291
Astragalus (genus containing locoweed species), 224
Atlantic bluefin tuna *(Thunnus thynnus),* 154–157, 325*n*44, 325*n*46
Atlantic bottlenose dolphin, 259–260
Atlas moth, 293
atmospheric carbon
bristlecone pines and, 112
redwood trees and, 48–49
Atomic Energy Commission, 58
Atropa belladonna (deadly nightshade), 226, 335*n*13
atropine, 226–227
aural sects, 159–187
how bats and moths are fighting evolutionary war, 178–181
howler monkeys, 159–162
how listening led to zoological revolution, 162–168
infrasound communication among elephants, 17
passive hearing in dolphins, 260
what crocodile sounds tell us about dino daddies and mommies, 172–174
what sperm whales tell us about listening to the world, 182–187
why all noisy creatures are not loudmouths, 169–172
why loudest frog changed its accent, 174–178
Austin, Christopher, 82–86, 170–171
Australian bulldog ant *(Myrmecia nigriceps),* 344*n*43
Australian tiger beetle *(Cicindela eburneola),* 146
aviation, 149–153
axolotls *(Ambystoma mexicanum),* 200–203, 332*n*20
Ayke, Kere, 13–15

B

Backster, Cleve, 288
bacteria
betaproteobacteria, 60–61
to consume radioactive waste, 59, 61–62
discovering, 67
diseases and, 59–61
healthy, 60
in human microbiome, 73–76
smallest, 67–68
spider genes transplanted into, 250
synthetic, 70–71
on tree of life, 63–64
Bacteria, 63
Bacteriodales, 75
Bailey (dolphin), 261
Baisan, Chris, 109
baji, 312*n*35
Balaenoptera musculus. see blue whale
Balaenopteridae, 35
baleen whales, 168
Banfield, Jillian, 63–64, 66–67
Bapan, China, 127–128
barbastelle bat *(Barbstelle barbastellus),* 180, 330*n*56
Bardo, Matt, 322*n*4
Barnes, Burton, 50–52, 76, 95–96, 297–299, 316*n*31
basilisk, 147
Bates, Paul, 88–90
Batista, Rafael Alves, 191, 192, 331*n*5
batrachotoxin, 234
bats, 87, 166. *see also specific types of bats*
calls of, 89–90
echolocation in, 184

in solving evolutionary mystery, 87–90
ultrasonic sounds of, 180–181
BBC, 179
Beale, Thomas, 182–183
Beard, Karen, 175–176
Beardsley, John W. "Jack," 76–77
beavers, 20, 105, 331*n*14
bebe (goliath frog), 4
beetles, 146, 208–212, 333*n*30
as food, 208
largest known, 208
smallest known, 93, 208
Begley, Sharon, 331*n*13
Bell, Alexander Graham, 328*n*33, 328*n*36
Benevides, Francis, 329*n*48
Bennett, Hayley, 315*n*16
Bertholdia trigona (tiger moth), 180, 330*n*55
betaproteobacteria, 60–61
Bhattacharya, Sanjib, 308*n*14
Bhugra, Dinesh, 267
biggest, 13–55. *see also* largest
discovery, forgetting, and re-discovery of largest plant, 50–55
goliath frog, 4–7
how elephants are like martial artists, 17–21
how tallest trees fight global warming, 46–49
how whale poop and terror attacks help in understanding stress, 40–46
humans' fascination with, 3–4
why blue whales are hard to research, 34–40
why elephant cells are like empathetic zombies, 21–26
why other animals benefit from elephant arousal, 26–29
why what we know about giraffes is mostly wrong, 30–34
biodiversity
in the Amazon, 213
and worlds slowest-evolving organism, 195–200
bio-inspired computing, 279–280
biological aging, 6
Biology Letters, 179
biomass, 48, 291–292, 345*n*59
Bioparque Del Amazonas, 213
bioweapons, 256
Bird, Christopher, 288
birds
aspen habitat for, 105
in Guam, 79
knowledge of dinosaurs based on, 173–174
smallest, 93
sounds made by, 172, 173
Birmingham, Amanda, 75
Bitcoin, 279
Black, Nancy, 35
black chokeberry *(Prunus virginiana melanocarpa),* 223
black mambas, 244, 245
blue whale *(Balaenoptera musculus),* 35–36, 65
brains of, 39
on California coast, 36
EBC samples from, 45
poop of, 42
population of, 36
senses used in choices made by, 39
size of aspens compared to, 51–52
in South Taranaki Bight, 37–40
territory of, 36
blue zones, 127
Bodnar, Andrea, 7
Bolt, Usain, 132
Boltzmann, Ludwig, 215
bone marrow recipients, 246, 247
Bonner, John, 19–20, 168, 310*n*11
Born Free Foundation, 131
Bothrops jararaca (Brazilian pit viper), 241
bottlenose dolphins, 157
Atlantic, 259–260
common, 262–263, 265
size and strength of, 264
speed of, 157
Bovidae, 142–143
bowhead whales, 115, 126
Boxall, Andy, 343*n*28
Boxin, 127–128
box jellyfish *(Chironex fleckeri),*

245–246
Boyle, Rebecca, 315*n*14, 343*n*29
Brachymyrmex, 280
Bradypus (three-toed sloth genus), 9, 204–206
brain
  of ants, 280
  of blue whales, 39
  cerebral cortex, 91–92
  of elephants, 266
  emotional control and, 264, 265
  of Etruscan shrew, 91–93
  of humans, 277, 279, 280
  IBM's Sequoia supercomputer simulation of, 279
  memory in, 290
  of octopuses, 274, 275
  of primates, reactions to snakes and, 236–238
  smallest, 8
  vision in, 92–93
Brazilian pit viper *(Bothrops jararaca)*, 241
Brazilian wandering spider *(Phoneutria fera)*, 249–250
Brenner, Sydney, 274–275, 342*n*18
Bridgeman, Laura, 264
Briggs, Helen, 322*n*6
bristlecone pine(s)
  death of, during scientific study, 118–119, 121
  growing at higher altitudes, 108–113
  Methuselah, 109–110
  negligible senescence in, 113–115
Bristol University, 216
British Medical Journal, 230
Britz, Ralf, 85
Brobdingnagians, 19–20
brown howler *(Alouatta guariba)*, 160
brown-throated sloths, 206
brown tree snake, 79, 80, 317*n*36
Brusatte, Stephen, 189, 330–331*n*1
Bryna, Jenny, 338*n*57
buffalo, 142
Bufo, 7
Buhler, Brendan, 316*n*27
bullet ant *(Paraponera clavata)*, 213–214, 307*n*3
bumblebee bat *(Craseonycteris thonglongyai)*, 87–90, 318*n*51
Bunn, Henry, 307*n*5
Burchell's army ant *(Eciton burchellii)*, 213
Burggren, Warren, 272
Burmese python, 235
Bush, Mitch, 138
Bushak, Lecia, 339–340*n*79
Byers, John, 144, 323*n*22, 323*n*23

C

Caenorhabditis elegans (roundworm), 221, 319*n*10
Callorhinchus milii (Australian ghost/elephant shark), 195–200
calls-and-balls study, 162
Canada, conifers in, 50
cancer, 310*n*18
  caused by tobacco, 228
  cembranoids for, 227
  cyclopamine for, 226
  development of, 24
  in dogs, 23
  in elephants, 24–26
  elephant shark and, 198
  hemlock for, 226
  leukemia, 247
  perforin use for, 246
  tobacco in fight against, 228–231, 335*n*20
captive breeding, 26–28
captopril, 241
carbon, global warming and, 48–49
carbon footprint, offsetting, 49
Carboniferous period, 174–175
Carnegie Mellon, 235
Cartagena Protocol on Biosafety, 256
castor bean *(Ricinus communis)*, 227
Castor canadensis (North American beaver), 331*n*14
Castoroides (giant beavers), 20
cats, 133, 134. *see also specific types of cats*
  histocompatibility of, 138–139
  Miracinonyx, 144
  speed of, 135–136
Cavener, Douglas, 311–312*n*30
CD4 gene, 199
cembranoids, 227–230

Center for Biological Diversity, 156
cephalopod-based artificial intelligence, 342*n*20
cephalopods, 274
cerebral cortex, 91–92
Cesarean births, 75, 76
cetaceans, limbic systems of, 264
Chandler, Todd, 38, 42
Charlton, Anne, 230
cheetahs *(Acinonyx jubatus)*
  speed of, 131–135, 142
  survival of, 137–141
Chetcuti, Nathan, 243
The Children of Men (James), 103
chimps, self-awareness in, 261
Chinese giant salamander, 9
Chinese river dolphin, 312*n*35
Chironex fleckeri (sea wasp/box jelly), 245–246
Choloepus (two-toed sloth genus), 205, 206
Christiansen, Paul, 314*n*53
Christie (elephant), 272
chromosomes, number of sets of, 100–102
chronic health problems, 59–60
Chronicle of Higher Education, 46
Chua, Leon, 343*n*25
Churchill, Winston, 79, 80
Church of the Holy Sepulchre, 64–65
Cicindela eburneola (Australian tiger beetle), 146
Cicuta douglasii (water hemlock), 224
"cigarette snails," 247
Clapper, James, 256
climate
  bristlecones as indicators of, 114
  sedimentary pollen fossils showing, 98–99
  tree ring indicators of, 111–112
climate change. *see also* global warming
  aspen die-offs and, 104
  coqui frog calls and, 178
  indicator animals for, 4, 85
  revealed in clam shell rings, 121
  seawater warming shown in hexactinellid spicule rings, 123, 124
  tree ring indicators of, 112
  trees' regulation of atmospheric carbon and, 48–49
  weather extremes with, 102
clonal organisms
  aspens, 51–55, 95–100
  as life-forms, 97
Clostridiales, 75
cnidarians, 245
cockroaches, 146, 323*n*27
cognitive maps, 284–285
Cohen, Michael, 320*n*27
Coleoptera, 209, 210, 249
collective intelligence, 283
Collembola (springtails), 249
Colombian bush cricket *(Arachnoscelis arachnoides),* 170
Comizzoli, Pierre, 29, 311*n*24
commercial activity, impact on stress levels of animals, 43
common bottlenose dolphin *(Tursiops truncatus),* 262–263, 265
common frog *(Rana temporaria),* 7
community, longevity and, 128–129
comparative genomics, 22–23, 115, 143
comparative oncology, 23–26
computers, 279–280
cone snail, 247–248, 339*n*74
conger eel, 40
congestive heart disease, 241
conifers, 50
Conium maculatum (true hemlock), 226
Conner, Steve, 342*n*17
Conraua goliath. see goliath frog
"Consider the Lobster" (Wallace), 210–211
Conus (genus of cone snails), 247
Conus magus (magical cone snail), 248
convergent evolution, 275
Cook, Dan, 223–225
Cooke, Lucy, 205–206, 332*n*24
Cooney, Michael, 189–190, 194–195
Coopersmith, Kaitlin, 324*n*31
Cope, Edward, 17, 144
"Cope's Cliff," 20, 21
Cope's Rule, 17–19, 84, 96
coqui frog *(Eleutherodactylus coqui),* 175–178
corals, 227, 245, 335*n*17

Correns, Karl, 180
Corson, Trevor, 333*n*35
The Cove (documentary), 263–264
cows, sounds made by, 165
Craseonycteridae, 87
Craseonycteris thonglongyai (bumblebee bat/Kitti's hog-nosed bat), 87–90, 318*n*51
Creature Encounters, 235
crickets, 170
criminal profiles, based on DNA, 66, 314*n*7
Crisanti, Andrea, 253
CRISPRCas9, 255, 256, 341*n*96
crocodiles, 173, 294
crocodilians
- knowledge of dinosaurs based on, 173–174
- parenting by, 173
- sounds made by, 172, 173

Crowder, Jim, 150
cryptocurrencies, 279
Currey, Donald, 118–120
Cusick, Kathleen, 132
cyanobacteria, 171
cyclopamine, 224–226
cyclopean lambs, 222–223
Cynoscion othonopterus (Gulf corvina), 294

D

Daintree Rainforest, 307*n*2
Daphnia pulex (water flea), 69
dart frogs, 232–233
Darwin, Charles, 30, 62, 342*n*18
Davis, Mark, 81–82
Davy, Philip, 117
Day, John, 127–128
deadliest, 219–257
- defining, 221
- how deadly reputation hamstrings world's most pharmaceutically promising plant, 227–231
- how deadly snakes made humans possible, 234–238
- how killer spiders and goats work together to make shoes, 248–251
- how poisons lead to cancer cures, 222–227
- why economic inequity stalled venom-based medicines, 239–244
- why largest pharmacy might be under the sea, 244–248
- why making mosquitoes less deadly could present bigger danger, 251–257
- why poisonous frogs don't poison themselves, 231–234

deadly nightshade *(Atropa belladonna),* 226, 335*n*13
Dean, Cornelia, 313*n*47
decision making, by plants, 291
deer, 105, 108
Delelegn, Demelash, 14–16
Dendroaspis polylepis (black mamba), 244, 245
dengue fever, 254
de Queiroz, Kevin, 312*n*33
de Vries, Hugo, 180, 332*n*18, 332*n*19
Dias, Brian, 342*n*13
Dictyostelium, 278
dinosaurs, 17. *see also specific dinosaurs*
- extinction of, 189
- possible sounds made by, 172–174
- speed of, 137
- understanding behavior of, 173–174

diplodocus, 307*n*4
Diptera (flies), 249
DiRenzo, Grace, 6, 7
Discover Magazine, 53
discovery, 294–300
disruption, interconnectedness and, 103–108
distress vocalizations, 327*n*14
distributed intelligence, 275–276
Di Ventra, Massimiliano, 278
DNA. *see also* genomes/genetic code
- of bluefin, 156
- creating composite sketches of persons from, 66, 314–315*n*7
- human, changes in, 96
- of humans and elephants, 22
- of *Populus tremuloides,* 54
- in prosecuting felons, 66

of *R. varieornatus,* 194
regulatory, 72
of smallest microbes, 63–64, 66–67
synthetic, 70–71
dogs, 166, 257
Dolbear, Amos, 328*n*33
Dolbear's Law, 170
dolphins. *see also specific dolphins by name; specific types of dolphins*
cognitive skills of, 259–263
conscious breathing by, 266
echolocation in, 184, 260
emotional control in, 264–266
limbic systems of, 264
non-aggression toward humans, 263–266
passive hearing in, 260
problem-solving by, 260
self-awareness of, 261
self-directed behavior in, 261–262
sounds made by, 165
Taiji slaughter of, 263–264
US Navy's use of, 156
Donath, Dirk, 334*n*1
Donoghue, Michael, 312*n*32
Dougherty, Elizabeth, 345*n*56
driving
listening to music while, 313*n*42
self-driving cars, 285–286
drones, 275–276
Drosophila melanogaster (fruit fly), 319*n*10
Drosophila subobscura (fruit fly), 319*n*10
Du Bois, Justin, 336*n*35
dung beetle, 210
Dye, Lee, 337*n*43
Dynastes hercules (Hercules beetle), 208

E

Earth Liberation Front, 47
Earth Microbiome Project, 59–60
earthworm, Gippsland, 8
eating, longest time without, 9
EBC (exhaled breath condensate), 44–46
Echis carinatus (saw-scaled viper), 244
echolocation
in bats, 180
in dolphins, 260
in sperm whales, 184, 185
US Navy studies of, 184
Eciton burchellii (Burchell's army ant), 213
Eciton hamatum (army ant), 283
The Ecology of Invasions by Animals and Plants (Elton), 79–80
economic inequity, 239–244
Ecuadorian mantled howler, 159–160
EDGE (evolutionarily distinct and globally endangered) species, 9, 16
Edible (Martin), 211
Elephant (journal), 16
elephants. *see also* African elephant; Asian elephant; *individual elephants by name*
brains of, 266
breeding and reproduction in, 26–28
cancer in, 24–26
captive breeding of, 268
in captivity, 342*n*11
cell divisions in, 24
evolution of, 18
hybridization of, 309*n*3
infrasound communication among, 17
killing of, 13–16
memory in, 266–267, 273
sounds made by, 163–166, 326*n*8
survival lessons from, 20–21
trauma and, 267–273
Elephants: Majestic Creatures of the Wild (Shoshani), 16, 309*n*5
elephant shark *(Callorhinchus milii),* 195–200
Elephas maximus. see Asian elephant
Eleutherodactylus coqui (coqui frog), 175–178
elk, 105
elm trees, 48
El Sayed, Khalid, 227–231
Elton, Charles, 79–80
emerald ash borer, 80
emotion, in dolphins, 264–265
emotional intelligence

in dolphins, 264–266
in elephants, 272–273
importance of, 265
empathy, 265, 272
endangered species
Atlantic bluefin tuna, 156
dart frogs, 233
goliath frog, 5–6, 308*n*9
gray whales, 41–42
Mexican walking fish, 201–202
preserving genetic diversity in, 29
preserving sperm of, 28, 29
on Red List of Threatened Species, 28
energy consumption. *see also* metabolic rate
by human versus synthetic brains, 279
by superorganisms, 216
Enserink, Martin, 340*n*89
Ensessakotteh wildlife refuge, 131–132
environment
change in animal calls and, 177–178
longevity and, 128–129
sonar in oceans, 185
sound and, 168
eogyrinus, 174
epibatidine, 232–233
epigenetics, 311*n*26, 342*n*14
defined, 271
trauma and, 271–273
erectile dysfunction, 249–250
Erisman, Brad, 345–346*n*1
Eschrichtius robustus. see gray whale
ethical standards, 256–257
Ethiopian Wildlife and Natural History Society, 14
Etruscan shrew *(Suncus etruscus)*, 91–93, 318*n*53, 318*n*54
Eukarya, 63
eukaryotes, 292
evolution, 310*n*15. *see also individual organisms*
bat-versus-moth in, 180–181
convergent, 275
Cope's Rule in, 17–19
Internet and, 217
Lamarckian inheritance, 30
role of stress in, 125
selective pressures in, 31–32
sensory drive hypothesis, 89–90
shared traits from, 292
size as driving force for, 19–20
slowest-evolving organism, 195–200
tree of life in, 62–65
using DNA to chart course of, 22
evolutionarily distinct and globally endangered (EDGE) species, 9, 16
Ewoks (tree-sitters), 47
exhaled breath condensate (EBC), 44–46
extinction, 5
of dinosaurs, 189
Holocene, 192, 233
intentional, 253–256
in Late Pleistocene, 137–138
recent scientific interest in organisms at risk for, 8–9
size and, 20
extreme phenomena, 8. *see also* superlatives

F

fairyflies *(Kikiki)*, 76–78
Falco peregrinus (peregrine falcon), 149–153
Falkowski, Paul, 73–74, 171
Fang, Janet, 253–254
fastest, 131–157
how bluefin tuna got into record books, 153–157
investigating, 296
why cheetahs should be but are not extinct, 137–141
why engineers are looking at falcons, 149–153
why mites are like Batman, 146–149
why pronghorns run from ghosts, 141–145
Fattah, Geoffrey, 340*n*81
fear
of snakes, 337*n*42
of spiders, 249, 337*n*42
Feltman, Rachel, 315*n*18
fertility, longevity and, 101–102

fire ant *(Solenopsis invicta),* 216–217
fires, around Fish Lake, 104
fish. *see also specific types of fish*
  killed and eaten by humans, 248
  noises of, 294–295
  smallest, 85
  venomous, 246
Fish Lake, 104
  aspen clone, 51–55, 95–100 (*see also* Pando [aspen colony])
  sediment from, 98–99
Flannery, Tim, 217, 334*n*42
Fleming, Alexander, 8
Flesher, Kristina, 298, 299, 346*n*2
flight, 149–153
Fontoura, Paulo, 193
food, longevity and, 128–129
food security, 207, 212
food sources
  for beetles, 209
  beetles as, 208–212
  for future humans, 209–210
  for sloths, 206
  of spiders, 248–249
Forbes, Andrew, 333*n*29
Fox, Stuart, 339*n*68
FOXO3 gene, 115–117
Franklin, Ken, 149–151
freshwater polyp *(Hydra vulgaris),* 116–117
freshwater snails, 257
Frightful (peregrine falcon), 150–152
frogs. *see also specific frogs*
  as indicator animals, 4, 85
  in learning about aging, 6–7
  measuring, 85
  metal pollutants impacts on, 308*n*8
  as model organism, 308*n*12
  most poisonous, 231–234
  smallest, 82–87
  used in research, 7–8
fruit fly
  Drosophila melanogaster, 319*n*10
  Drosophila subobscura (fruit fly), 319*n*10
Fry, Bryan, 245–246, 339*n*69
funding for research, 45–46
fungus, largest, 53, 297
Furlong, Michael, 332*n*16

G

Gagliano, Monica, 290–291, 345*n*54
Gaile, 317*n*34
Galleria mellonella (greater wax moth), 178–181
Galton, Francis, 166, 327*n*18
garden pea *(Pisum sativum),* 290–291
Gates, Bill, 251, 254, 340*n*82
Geirland, John, 309*n*6
gender, lightening deaths and, 334*n*4
gene-drive-altered animals, 254–256
General Sherman (sequoia), 52
genetically modified insects, 254
genetically modified mosquitoes, 81, 254–257
genetic diversity
  of aspens, 97–98
  in giraffes, 32–33
  rare human genetic conditions and, 99
  in saving endangered animals, 28
Genome 10K project, 197
genome resource banks, 29
genomes/genetic code
  of axolotls, 201
  of bowhead whales, 126
  of bumblebee bats, 90
  of cheetahs, 138–140
  of coelacanth, 331*n*11
  comparative genomics, 22–23, 115, 143
  of elephants, 22
  of elephant sharks, 197–201
  of giraffes, 31–32, 311–312*n*30
  of humans, 22, 71, 331*n*13
  of hydra, 116
  of JCVI-syn3.0, 71
  length and complexity of, 69–70
  longest, 200–203
  longevity and, 99, 128–129
  of *Mycoplasma genitalium,* 70
  of *Mycoplasma mycoides,* 70–71
  of octopuses, 275, 342*n*17
  of okapi, 31–32
  of Pando, 96
  of *Populus tremuloides,* 125
  of *R. varieornatus,* 194
  for smallest organisms, 63–64, 67
  triploids, 100–103

of water flea, 69
widespread inbreeding and, 137–139
genomic research, evolutionary links offered by, 63
genomic technologies, actionable ethical standards for, 256–257
Ghiselin, Michael T., 311*n*25
GhostSwimmer, 156–157
giant armadillo *(Glyptodon),* 20
giant beavers *(Castoroides),* 20
giant panda, 206
Gibson, Dan, 70
Gibson, Prudence, 292, 345*n*60
Gigantopithecus (ape), 20, 310*n*13
Gilbert, Jack, 59–60
Gillooly, Jamie, 215
Gippsland earthworm, 8
giraffes
classified as single species, 32–34
conservation of, 33–34
evolution of, 30–33, 143
genetic diversity in, 32–33
genome sequence for, 311–312*n*30
sounds made by, 166
glass sponge *(Monorhaphis chuni),* 122–125
Gliese 445, 10
global warming. *see also* climate change
Bitcoin electricity use and, 279
micro-frogs as indicators of, 85
redwood trees' regulation of atmospheric carbon, 48–49
tree ring indicators of, 112
Glyptodon (giant armadillos), 20
goats, 250–251
Godfrey-Smith, Peter, 274, 342*n*16
Goldberg, Rube, 72
golden howler *(Alouatta palliata),* 160
golden poison dart frog, 232
goliath frog *(Conraua goliath),* 4–7, 306–307*n*6, 308*n*9
Goode, 317*n*40
Goodman, Morris, 22, 310*n*16
Google, 281–282
Gore, Al, 340*n*81
Gorli syndrome, 226
Grand Cayman island, 254
Grant, Michael, 52–53
grant applications, 88
grassy deathcamas *(Zigadenus gramineus),* 223
gray whale *(Eschrichtius robustus),* 36, 40–42, 312*n*35
collecting poop samples from, 42–44
removal from endangered list, 42
greasewood *(Sarcobatus vermiculatus),* 223
great apes, 334*n*41
greater wax moth *(Galleria mellonella),* 178–181
Great Oxygenation Event, 171
Greenwood, Faine, 325*n*44
Grevillea renwickiana, 101–103, 125, 319*n*9
Griffin, Donald, 166
Grollier, Julie, 279–280, 343*n*31
Guam bird population, 79
Guarino, Ben, 331*n*4
Guarino, Emily, 260
Guinee, Linda, 326*n*6
The Guinness Book of World Records, 3, 146
Gulf corvina *(Cynoscion othonopterus),* 294
Gulliver's Travels (Swift), 19

H

Hansen, Slim, 118
Harkes, John, 329*n*47
Harlan, Tom, 109–110, 320*n*17
Harpole, Tom, 150, 325*n*35
Harvard University, 277
Hawai'i, coqui frogs in, 175–177
Hawking, Stephen, 191, 331*n*3
Haxel, Joe, 43
health technologies, actionable ethical standards for, 256–257
Heath, David, 335*n*20
Hemiptera, 249
Hendrickson, Susan, 315*n*8
Hercules beetle *(Dynastes hercules),* 208
Herper, Matthew, 334*n*10
Herzing, Denise, 265
hexactinellids, 122–125

Hickman, Tom, 326*n*4
Higley, Brewster, 143
Hiltzik, Michael, 313*n*48
hippos, sounds made by, 166
Hirt, Myriam, 135
histiocytosis, 23
histocompatibility, 138–139, 160
Hoehl, Stephanie, 337*n*44
Hoffman's sloths, 206
Hogle Zoo, 23
Holocene Extinction, 192, 233
hominoids, 4
Honolulu-Asia Aging Study, 320*n*25
Hopkins, Chris, 247–248
Howard, Jacqueline, 334*n*2
howler monkeys, 159–163
Hubbard, Daniel, 321*n*9
Huber, John, 76–78
Hug, Laura, 64
Hughes, Malcom, 320*n*19
humans
- aging in, 6–7
- brain's response to snake images in, 238
- causes of death for, 334*n*2
- common genes of other organisms and, 21–22
- compensation cars of, 162
- depression in, 165
- DNA changes in, 96
- dolphins' imitation of, 260
- dolphins' non-aggression toward, 263–266
- emotional self-control in, 265–266
- energy needed for brain in, 279
- fear of snakes in, 235–236
- fear of spiders in, 249
- genomes of, 22, 71, 331*n*13
- Holocene Extinction, 192, 233
- immune system of, 60
- killed by box jellyfish, 245
- killed by dogs, 257
- killed by freshwater snails, 257
- killed by mosquitoes, 221, 252
- killed by snakes, 235, 236, 240–241, 257
- longevity factors for, 127–128
- meat and fish killed and eaten by, 248
- memory in, 279, 290
- microbiome of, 73–76
- navigation by, 284–285
- people killed by, 221
- regeneration in, 201
- self-awareness of, 261
- shared gene with elephant shark, 198
- social evolution of, 217
- understanding potential of, 10–11

Humberto Madrid, John, 212–214, 217
Humboldt State University, 48–49
humpback whales, 35, 36, 163, 326*n*6
hunting, 307*n*5
hydra, 116
Hydra oligactis, 320*n*24
Hydra vulgaris (freshwater polyp), 116–117
hyenas, 323*n*23
hygiene hypothesis, 60
Hymenoptera, 249
Hyperion (redwood), 48, 52, 296
hypertension, 241

I

IBM, 279
Icaronycteris (extinct bat genus), 180
immune system
- of elephant sharks, 199–200
- of humans, 60, 199

The Independent, 120–121
indicator animals
- for climate change, 4
- for global warming, 85

industrial activity, impact on stress levels of animals, 43
infrasound, 17, 163–166, 326*n*8
Ingraham, Christopher, 248–249
inheritance, 311*n*25
- epigenetic, 271–273, 311*n*26, 342*n*14
- Lamarckian, 30
- Mendelian, 30

Inky (octopus), 273
inland taipan *(Osyuranus microlepidotus),* 242
insects. *see also specific insects*

as food source, 208–212
population of, 333*n*33
Tinkerbella nana, 78, 87
Institute of Zoology, 308*n*12
Intel, 276
intelligence, 263. *see also* artificial intelligence; smartest
of ants, 283–285
associative learning in, 291
collective, 283
distributed, 275–276
emotional, 264–266, 272–273
evolution of, 274
factors in, 39
measuring, 283
intelligence collective, 283
intentional extinction, 253–256
interconnectedness, 103–108
International Union for Conservation of Nature and Natural Resources (IUCN), 5, 28, 34, 156
International Whaling Commission, 35
Internet
evolution and, 217
search engines, 281–282
interplanetary space travel, 194
invasive species, 78–82, 175–176
invertebrate, smartest, 273
Irwin, Steve, 246
Isabela (whale), 312*n*40
Isbell, Lynne, 30–31, 236–238, 311*n*28, 337*n*47
Istiophorus platypterus (sailfish), 153–155
IUCN. *see* International Union for Conservation of Nature and Natural Resources

J

Jaakkola, Kelly, 262–263
Jackson, Alan, 43
James, C. Renée, 339*n*73
James, Elizabeth, 101–103
James, P. D., 103
Janke, Axel, 32, 33
Janssen, Antti, 169
Jax (dolphin), 262
JCVI-syn3.0, 71
jellyfish, 244–246
Jenkins, Steve, 293–294, 306–307*n*6
Jenny (Asian elephant), 266–267
Jerusalem, 64–65
Jesus lizard, 147
Jinka, Rico, 220
Jochum, Klaus, 122–123
Johns Hopkins, 225, 230
Johnson, Andrew, 332*n*20
Jørgensen, Aslak, 193
Journal of the Kansas Entomological Society, 179
june beetle, 210
Jurassic Park (movie), 136–137

K

Kaplan, Sarah, 339*n*72
Kefelioglu, Haluk, 318*n*53
Kelly, Morgan, 344*n*41
Kerr, Iain, 44–45
Ketten, Darlene, 326–327*n*12
Kickstarter, 45
kidney problems, 241
Kikiki (genus of fairyflies), 76
Kikiki huna (species of fairyfly), 77–78
killers. *see* deadliest
King, Denny, 100–101
King Kong, 3
King's holly *(Lomatia tasmanica),* 100–102, 125, 319*n*7
Kinzley, Colleen, 27
Kipchoge, Eliud, 132
Kirby, Eric, 36
Kiso, Wendy, 28, 29
Kitti's hog-nosed bat *(Craseonycteris thonglongyai),* 87–90
Knapp, Leslie, 160–162, 326*n*2
Knight, Rob, 59–60, 74–75, 316*n*24
Kobayashi, Ryo, 277
Kodjak, Alison, 335*n*21
Koetsier, John, 343*n*35
Koren, Marina, 309*n*17
Krulwich, Robert, 345*n*55
Krywko, Jacek, 343*n*27
Kunieda, Takekazu, 194, 331*n*9
Kurths, Jurgen, 282, 283

L

Lamarck, Jean-Baptiste, 30, 311*n*25

Lamarckian inheritance, 30
Lambert, Darwin, 118–119
largest. *see also* biggest
amphibian, 9
aspen colony, 50–55
earthworm, 8
fungus, 53
Late Pleistocene extinction, 137
La Trobe University, 230–231
Lawler, Susan, 231
Laws, Richard, 16–17
Learn, Joshua Rapp, 341*n*4
learning, 290–291
associative, 291
machine, 343*n*32
leather corals, 335*n*17
Leclerc, Georges-Louis, Comte de Buffon, 205
Le Conte, John Eatton, 333*n*30
Lecoz, Abbe, 330*n*57
Ledecky, Katie, 132
Leiden University, 239
lemurs, 8, 237
Lennon, Jay, 65–66
Leopold, Aldo, 108
Lepidoptera, 249
lesser water boatman *(Micronecta scholtzi),* 169
leukemia, 247
Lewis, Randy, 250
life, new model for, 203
Life's Engines (Falkowski), 171
Li–Fraumeni syndrome, 24
Lilienthal, Otto, 153
limbic systems, of cetaceans, 264
Lin, Sarah, 329*n*42
Lindstedt, Stan, 145
Linnaeus, Carolus (Carl), 32–33
lions, humans killed by, 221
lion's mane jellyfish, 246
living things, human connections to, 10
lizards, 84, 147
Lloyd, Graham, 319*n*6
lobsters, 210–211
Locey, Ken, 65–66
locoweed, 224
Lomatia tasmanica (King's holly), 100–102, 125, 319*n*7
longest-lived. *see also* oldest
goliath frog, 6
mammals, 115
trees, 48
longevity, 99
of bowhead whales, 115
of bristlecone pines, 113–115
how trees, whales, and polyps help us live longer, 113–117
for humans, 127–128
of *Hydra vulgaris,* 116–117
reproduction and, 101–102
simplicity-stress-survivability equation for, 125–129
stem cells and, 125
sterility and, 100–103
what sponges, trees, and whales teach about longevity, 124–129
Long Key Fishing Camp, 153
the Lorax (tree-sitter), 47
loudest. *see also* aural sects
frog, 174–178
investigating, 295–296
Lova, Bulisa, 83
Loxodonta africana. see African bush elephant
Loxodonta cyclotis (African forest elephant), 309*n*2
Lü, Junchang, 189
Luers, Jeffrey "Free," 47
Lutcavage, Molly, 154–155, 325*n*42
Lutz, Matthew, 283

M

machine learning, 343*n*32
Madagascar ragwort *(Senecio madagascariensis),* 225
Maffly, Brian, 334*n*10
magical cone snail *(Conus magus),* 248
magnetic navigation system, of sperm whales, 185–186
Mago National Park, 13–16
Makhloufi, Taoufik, 132
malaria, 221, 251, 253
antimalarial drug resistance, 340*n*85
spending on control of, 340*n*84
Maloney, Jennifer, 335*n*22, 335*n*23
mammals. *see also specific mammals*
size of, 310*n*12

slowest, 204–207, 332*n*23
sounds made by, 167–168
speed of, 136
mammoths, 18, 137–138
Mann, John, 335*n*13
Marcot, Jonathan, 209
Marí, Frank, 247
marine mammals, sounds made by, 167–168
Maron, Dina Fine, 310*n*18
Marshall, Aarian, 344*n*48
Marshall, Craig "Critter," 47, 313*n*49
Marshall, John, 341*n*97
Martin, Daniella, 211
Martin, Kobe, 166–168
Martín, Piero, 204
Martínez, Daniel, 114–117
"Martyr for a Species" (Lambert), 118–119
Masai giraffe, 33
Mashburn, Kendall, 45
Massachusetts Institute of Technology, 156, 277
Matao, 127
mating traits, sensory drive hypothesis and, 89–90
Mautz, William, 329*n*48
Max Planck Institutes, 235–236
Mayo Clinic, 230
Mayr, Ernst, 33
McAvoy, Darren, 319*n*12
McCardel, Chloe, 132
McComb, Karen, 269
McKie, Robin, 307*n*5
McKinnon, Shaun, 319–320*n*16
McKnight Brain Institute, 113–114
McMullen, Christina, 260
McNulty, Dan, 105
mealworms, 208
Medunda (elephant), 310–311*n*20
megalocephalus, 174
Megalonyx (ground sloth), 20
Mellisuga helenae (bird), 93
Melville, Herman, 186
memory
in ants, 284, 285
artificial, 279
in elephants, 266–267, 273
in humans, 279, 290
in single-celled organisms, 278
of trauma, 267–273
memristors, 278, 280, 343*n*25
Mendel, Gregor, 179–180
Mendelian inheritance, 30
meranti, 296
meristematic cells, 126
Merrit, Thomas, 146
metabolic rate
of ant colonies, 215–216
lowest, 9
of three-toed sloths, 206
Metabolic Scaling Theory, 215
Methuselah (bristlecone pine), 109–110
Mexican walking fish *(Ambystoma mexicanum),* 200–203
Miaud, Claude, 5–7
mice
aging of, 7
trauma experiments on, 271–272
microbes
bacteria, 59–62, 67–68
to consume radioactive waste, 58–59, 61–62
discovering, 65–69
human microbiome, 73–76
science's ignoring of, 73–74
on tree of life, 63–64
Microcebus berthae (primate), 93
micro-frogs, 82–87
Micronecta scholtzi (lesser water boatman), 169
microscope, 73
migration
of aspen colonies, 52, 98–99
of bristlecone pines, 112
change in animal calls and, 177
of nonnative species, 78–82
of whales, 312*n*40
Miller, Patrick, 184
mimic octopus *(Thaumoctopus mimicus),* 273
Mimosa pudica (touch-me-not plant), 290
mindset, longevity and, 128–129
Ming (quahog clam), 120–121
Miracinonyx inexpectatus (false cheetah), 144
Miracinonyx trumani (false cheetah), 144

mites *(Paratarsotomus macropalpis),* 146–149, 295
Moalem, Sharon, 319*n*6
Mock, Karen, 54, 100
Mohanraj, Prashanth, 77
mollusk, smallest, 93–94
monkeys, 159–163, 237–238
Monorhaphis chuni (glass sponge), 122–125
Morelle, Rebecca, 321*n*31
Morris, Jesse, 98–99
Morrison, Rachel, 261–262
Moscow State University, 102
mosquitoes, 251–257
genetically modified, 81, 254–257
humans killed by, 221, 252
intentional elimination of, 253–256
tiger mosquito, 79, 81, 317*n*38
moths, 178–181, 293, 330*n*55
motion, longevity and, 128–129
Motty (hybrid African–Asian elephant), 309*n*3
mountain lions, 105
Müller, Werner, 321*n*34
Munné-Bosch, Sergi, 126, 321*n*35
mutations, to accommodate growth, 19–21
Mycoplasma genitalium (parasitic bacterium; possibly the smallest living organism), 68, 70
Mycoplasma mycoides, 70–71
Mymaridae (family of wasps), 76
Myrmecia nigriceps (Australian bulldog ant), 344*n*43
"My Western Home" (Higley), 143

N

NaD1, 231
Nagar, Apoorva, 285
Nakagaki, Toshiyuki, 277, 278
Nanoachaeum equitans (extremophilic archaeon), 67–68
Narcissa (snake), 235
Narins, Peter, 176–178
National Geographic, 179
National Institutes of Health, 230
National Wildlife Federation, 81
Nature, 53
Naumann, Robert, 92, 318*n*54
navigation
by ants, 283–287
self-driving cars, 285–286
Navy Marine Mammal Program, 156
negligible aging, 320*n*23
negligible senescence, 113–117
nematocysts, 245
Neotrogla, 342*n*21
neurobiology, plant, 289–292
neutrinos, 10
New England Aquarium, 43
New York Times, 179
New Zealand Environmental Protection Authority, 39
Nicholls, Henry, 311*n*27
Nicotiana tabacum. see tobacco
nicotine, 228
Niedernhofer, Laura, 6, 7
Nobel Prize ceremony/lectures, 276–277, 338*n*59
nonnative species, migration of, 78–82
North American beaver *(Castor canadensis),* 331*n*14
northern giraffe, 33, 34
North Korea, 308*n*16
Novacek, Michael, 330*n*54
Noyes, John, 78
Nuwer, Rachel, 340*n*80
Nye, Bill, 309*n*18

O

Oakland Zoo, 26
oak trees, 48
object permanence, 262
O'Brien, Stephen, 138, 140–141, 197, 322*n*13
Ocean Alliance, 44, 45
octopuses, 273–276, 342*n*17
Odelwald, Sten, 313*n*46
okapi, 31–32, 311–312*n*30
OK Go, 72, 315*n*19
Okinawa, centenarians in, 127
Okinawa Trough, 122, 123, 125
oldest, 95–129. *see also* longevity
deaths of very old organisms and scientists' study of superlatives, 118–121
how aspens teach about

interconnectedness, 103–108
how oldest known animals helps unlock ocean's secrets, 122–124
how sterility helps organisms grow old, 100–103
how trees, whales, and polyps help us live longer, 113–117
humans, 127
Pando aspen colony, 95–100
what sponges, trees, and whales teach about longevity, 124–129
why bristlecone pines are growing at higher altitudes, 108–113
O'Leary, Maureen, 309*n*9
Olivera, Baldomero, 248
olm, 9
On the Origin of Species (Darwin), 30, 62
ophiderpeton, 174
ophidiophobia, 235
Orthoptera, 249
Osh (bull elephant), 26–27, 310–311*n*20
ostracod crustaceans, 315*n*15
ostrich, 293–294
Other Minds (Godfrey-Smith), 274
outliers, 8
Owen, Nisha, 16
Oxitec, 254, 256, 340*n*88, 340*n*89
oxygen
in atmosphere, 171
Great Oxygenation Event, 171
oxygen transport
in cheetahs, 135–136
in pronghorns, 145, 323*n*18
Oxytropis (genus containing locoweed species), 224
Oxyuranus microlepidotus (inland taipan), 242

P

p53 gene, 24–25, 198, 226
Pacific sea nettles, 246
Paedocypris progenetica (SE Asian fish), 85
Paedophryne amauensis (world's smallest frog), 7, 82–87
Pagán, Oné R., 335*n*18
Paget, James, 99
painkillers, 243, 248
Palaeochiropteryx (extinct bat genus), 180
palaeomastodon, 18
Pancake (gray whale), 40–44
Pando (aspen colony), 296–297
age of, 95–100
disruption and possible death of, 103–107
longevity factors for, 125–126
size of, 53–55
sterility of, 100
Pangu, 3
Panthera, 327*n*38
Papua New Guinea, 82–85
Paraponera clavata (bullet ant), 213–214, 307*n*3
Paratarsotomus macropalpis (mite), 146–149, 295
parenting
by crocodilians, 173
by poisonous frogs, 231
pattern recognition, 280
Paul, Jason, 324*n*37
Pauli, Jonathan, 206
Pauwels, Eleonore, 255–256
PAX3 gene, 202
PAX7 gene, 202
Payne, Katherine, 163–165, 326*n*6
PEEL Therapeutics, 310*n*19
PellePharm, 225–226
PELskin project, 325*n*38
penicillin, 8, 60
penicillin allergies, 308*n*14
Penn State University, 31–32, 66
Penny, David, 311*n*26
peregrine falcon *(Falco peregrinus),* 149–153
perforin, 246–247
Periplaneta americana (American cockroach), 146, 323*n*27
Perkins, Sid, 321*n*7
Peterson, Eric, 23–24
Peto, Richard, 23
Peto's Paradox, 23
Pfister, Jim, 224, 225
Phillips, Tom, 312*n*35
Phiomia, 18
Phoneutria fera (Brazilian wandering

spider), 249–250
phosphatherium, 18
photoacoustic effect, 328*n*36
photosynthesis, 171
Phyllobates terribilis (dart frog), 232
Physarum polycephalum. see slime mold
Physeter macrocephalus. see sperm whale
phytoremediation, 335*n*15
Pilanesberg Park, 269, 270
Pinus ponderosa (Ponderosa pine), 225
pistol shrimp, 170, 328*n*29, 328*n*31
Pisum sativum (garden pea), 290–291
Plait, Phil, 321*n*9
Plantae, 48
plants
- information collection/processing, and sharing by, 288–289
- neurobiology of, 289–292
- singing to, 287, 288

Plasmodium (malaria parasites: *P. falciparum, vivax, ovale, malariae,* and *knowlesi*), 252
Plasmodium falciparum, 252–253
Pliny the Elder, 68–69
PnTx2-6, 250
poisonous animals
- frogs, 231–234
- in the oceans, 244–248
- snakes, 234–237, 339*n*66
- spiders, 248–250
- venom-based medicines, 239–244
- venomous species, 239

Poisonous Plant Research Laboratory, 223
poisonous plants, 223–227
- for soil decontamination, 227
- types of poisoning, 224–225
- used in medicine, 226–231

Pollan, Michael, 345*n*53
pollen fossils, 98–99
Polyphaga, 333*n*32
Polysphondylium, 278
Ponderosa pine *(Pinus ponderosa),* 225
Poole, Joy, 326*n*9
poop
- human, in microbiome research, 75
- whale, 42–44

population bottleneck, 138, 140–141, 322*n*15
populations
- of axolotls, 202
- of blue whales, 36
- of bumblebee bats, 318*n*51
- of cheetahs, 139
- of elephants, 14–15, 268, 309*n*1
- of humans, 47
- of insects, 333*n*33
- of lions, 221
- of trees, 48, 313*n*50
- of whales, 186–187

Populus tremuloides. see quaking aspen
post-traumatic stress disorder (PTSD), 267–268, 270, 341–342*n*10
Poulain, Michael, 127
Prasinohaema (Papua New Guinean lizards), 84
predators
- cheetahs, 139
- ecosystem role of, 105–108
- of pronghorns, 143–144
- wolves, 105, 108

Prescott, Tony, 92–93
primates. *see also specific types of primates*
- histocompatibility genes in, 160
- smallest, 93
- snakes and, 236–237

Princeton University, 277–278
problem-solving
- by dolphins, 260
- by humans, 277
- by octopuses, 273–276
- by single-cell slime molds, 277–278, 280

proboscideans, 326*n*11
Proceedings of the Hawai'ian Entomological Society, 77
Procopius, 111
Prokic, Marko, 308*n*8
Prometheus (bristlecone pine), 118–119, 121
pronghorn *(Antilocapra americana),* 141–145, 323*n*18
protection, 15–16
- of aspen clone, 53, 106–107
- of blue whales, 35, 37, 39–40

of Mexican walking fish, 202
of oldest trees, 110
of rainforests, 233
Prunus virginiana melanocarpa (black chokeberry), 223
Ptilium fungi (beetle species), 333*n*30
PTSD. *see* post-traumatic stress disorder
public relations for research, 86–87
Pugh, Jonathan, 256, 340*n*93
purpose, longevity and, 128–129
pygmy mouse lemur, 8

Q

quagga mussel, 79, 81
quahog clam *(Arctica islandica)*, 120–121
quaking aspen *(Populus tremuloides)*, 50–51, 54, 102, 125, 298. *see also* Pando (aspen colony)
Quong, Pang, 197

R

radioactive waste, 58–59, 61–62
radiocarbon dating, 114
Raimer, Henrik, 132
Rainbow (bottlenose dolphin), 259, 260
rainforest, 233
Raloff, Janet, 329*n*43
Ramazzottius varieornatus (tardigrade species), 193–194
Rana, 7
Rana temporaria (common frog), 7
rats, sounds made by, 166
Raytheon, 342*n*20
RBP-J gene, 90
Red List of Threatened Species, 28
redundancy, 203
Redwood Forest Foundation, 49
redwoods, 48–49, 310*n*14, 314*n*52
Reed, Robert, 341*n*95
regeneration, 201
Reiss, Diana, 261
release of insects with dominant lethality (RIDL), 81
Remsen, David, 323*n*20
Rensch, Bernhard, 309*n*8
reproduction
longevity and, 101–102
of sloths, 206–207, 332*n*27
of triploids, 100, 101
reproductive research, 28–29
reproductive status, studying EBC to determine, 45
research funding, 45–46
Ressler, Kerry, 342*n*13
reticulated giraffe, 33
Rettner, Rachael, 333*n*37
rhinoceros beetle, 210
rhinos, sounds made by, 166
Rhody (Schiffman's dog), 23
rhythm, longevity and, 128–129
ricin, 227
Ricinus communis (castor bean), 227
RIDL (release of insects with dominant lethality), 81
Rifle Integrated Field Research Challenge, 57–59, 61–68, 93–94
Ringling Bros. Center for Elephant Conservation, 28, 29
Ritchie, James, 341*n*9
Rittmeyer, Eric, 83
robots, 92–93, 156–157
RoboTuna project, 156
rock hyrax, 63
rodents. *see also specific rodents*
communication of distress in, 327*n*17
sounds made by, 165, 166
Roe, Paul, 308*n*7
Rogers, Paul, 103–107, 297–298
Rogers, Tracey, 327*n*22
Rojas, Shirley Jennifer Serrano, 233
Rosenberg, Robin, 324*n*33
Rosti, Marco, 153–153
roundworms *(Caenorhabditis elegans)*, 221, 319*n*10
rover ants, 280
Rowell, Timothy, 345–346*n*1
Royal Society of London for Improving Natural Knowledge, 73
Rubin, Samuel, 146–148
Rutgers, 235

S

Sage (tree-sitter), 47
sailfish *(Istiophorus platypterus)*,

153–155
Salomons, Hannah, 341*n*2
saltwater crocodile, 294
Salzer, Matthew, 110–112, 119
Sarah (cat), 133
Sarcobatus vermiculatus (greasewood), 223
Satel, Sally, 336*n*26
Save the Redwoods League, 49
saw-scaled viper *(Echis carinatus),* 244
scaling, 148–149
Schiffman, Josh, 22–26, 99
Schipani, Sam, 332*n*21
schistosomiasis, 257
Schmidt, Justin, 214, 307*n*3
Schmidt Pain Scale, 214
Schobert, Les, 268
Schroeder, Avi, 26
ScienceDirect database, 4, 8
scientific ambassadors, 9
Scourse, James, 121
Scydosella musawasensis (beetle), 93, 208
sea lions, US Navy's use of, 156–157
search-and-rescue operations, 276
search engines, 281–282
Search with Aerial RC Multirotor (SWARM), 276
sea urchins, non-aging of, 7
sea wasp *(Chironex fleckeri),* 245–246
The Secret Life of Plants, 287–288
Selcuk, Ahmet, 318*n*53
self-awareness, in dolphins, humans, and chimps, 261
self-directed behavior, in dolphins, 261–262
self-driving cars, 285–286
Senecio madagascariensis (Madagascar ragwort), 225
senescence, 6, 113
- in frogs, 6–7
- negligible or lack of, 113–117

Sensenig, Kate, 329*n*41
sensory drive hypothesis, 89–90
Senter, Phil, 327*n*37
Seong, Jae Young, 332*n*16
Seoul National University, 325*n*43
Sepkoski Curve, 66
September 11, 2001 terrorist attacks, 43
Sequoia supercomputer, 279
sharks
- genomes of, 198
- human deaths from, 222

Shaw, George, 201
sheep, 223–224
Shirley (elephant), 267
Shoshani, Jeheskel, 16, 309*n*4
Shrewbot, 93
Shriver, Mark, 66
Shubin, Neil, 196, 331*n*10
Silent Nemo, 156
Siler, Wes, 315–316*n*20
silkworms, 250
simplicity-stress-survivability equation, 125–129
Sinclair, David, 115–116
Singer, P. W., 326*n*8
Singer, Sydney Ross, 176
single-celled organisms
- as key to artificial intelligence, 277–278, 280
- memory in, 278

Sinularia gardineri (soft coral), 227
Sinularia species, 335*n*17
size. *see also* biggest; largest; smallest
- age and, 110
- of aspens, 98 (*see also* Pando [aspen colony])
- calling frequency and, 166–168
- cancer rates and, 23
- comparative, 10
- "Cope's Cliff" and, 20, 21
- Cope's Rule for, 17–19, 84, 96
- deadliness and, 221
- energy expended and, 215
- genome length and, 69, 70
- of Hyperion, 52
- intelligence and, 280
- of mammals, 310*n*12
- mutations producing, 18
- speed and, 135–136, 146, 148–149

Skoch, Iva Roze, 332*n*27
Skye (tree-sitter), 47
slime mold *(Physarum polycephalum),* 277–278, 280, 342–343*n*22
Sloan, David, 191, 192
sloths, 9, 20, 204–207, 332*n*27
slowest mammal, 204–207, 332*n*23

smallest, 57–94
  brain, of pygmy mouse lemur, 8
  how bat helped solve evolutionary mystery, 87–90
  how Etruscan shrew helps understand brain, 90–94
  how microbes helped draw new tree of life, 62–65
  how smallest flying animal belies invasive species beliefs, 76–72
  how smallest life-forms help solve genetic code, 69–72
  how smallest vertebrate helps understand past climates, 82–87
  why microbiologists are like criminal prosecutors, 65–69
  why microbiomes in humans may be as important as brain, 73–76
smartest, 259–292
  how ants teach us to navigate complex world, 280–287
  how elephant memories help us understand trauma, 266–273
  how octopuses are like a football team, 273–276
  what we learn from smarty plants, 287–292
  why dolphins don't kill us, 263–266
  why single-celled organisms are key to artificial intelligence, 276–280
Smith, Dena, 209
Smith, Manard, 319*n*10
Smokey, 2
smoking, 228–229
snails, 93–94, 247–248, 257, 339*n*74
Snake Detection Theory, 236
snakes, 79, 80, 317*n*36
  attentiveness to, 338*n*52
  fear of, 235–236, 337*n*42
  human deaths from, 235, 236, 240–241, 257
  odds of being killed by, 222
  primates' reactions to, 236–238
  species of, 339*n*66
  venomous, 234–235, 242–244, 339*n*66
SnotBot, 45
Snyder, Laura, 316*n*22
soft coral *(Sinularia gardineri),* 227
soil decontamination, 227
solar storms, 186
Solenopsis invicta (red imported fire ant), 216–217
sound(s). *see also* aural sects
  of archosaurs, 172–173
  in Carboniferous period, 174–175
  as drivers of speciation, 161
  first, by any life-form, 171
  infrasonic, 17, 163–166, 326*n*8
  made by elephants, 163–166
  outside of human range of hearing, 163
  of sperm whales, 183
  ultrasonic, 166–167, 170, 178–181
South American black howler *(Alouatta caraya),* 160
southern giraffe, 33
South Omo River Valley, 14
South Taranaki Bight, 37–40
space travel, 194
Spangler, Hayward, 178–179, 329*n*51
speciation, 33
species
  of amphibians, 336*n*32
  of ants, 214
  of beetles, 208–209
  of giraffes, 32–34, 311–312*n*30
  of insects, 333*n*33
  invasive, 78–82
  Linnaeus' classification of, 32–33
  of mosquitoes, 253
  number of, 65–66, 309*n*10
  of plants, 291
  of proboscideans, 326*n*11
  scientific definition of, 33
  of snakes, 339*n*66
  of spiders, 249
  venomous, 239
  of venomous snakes, 244
  Woese's classification of, 62–63
spectacled caiman, 173
speed. *see also* fastest
  of bluefin tuna, 155
  of bottlenose dolphin, 157
  of cats, 135–136

of cheetahs, 131–135, 142
computing, in humans and simulated brains, 279
different measures of, 132
of dinosaurs, 137
of falcons, 150
of mites, 147
of pronghorns, 142–144
of robot fish, 157
size and, 135–136, 146, 148–149
slowest mammal, 204–207
of universe expansion, 309*n*17
spermaceti, 183
sperm whale *(Physeter macrocephalus)*, 182–187
echolocation in, 184, 185
magnetic navigation system of, 185–186
Sphaerodactylus ariasae (reptile), 93
spiders, 221–222, 248
animals killed and eaten by, 248–249
fear of, 249, 337*n*42
silk from, 250–251
spider silk, 250–251
sponges
longevity factors for, 124–125
Monorhaphis chuni, 122–125
springtails (Collembola), 249
Staedter, Tracy, 342*n*20
Steinhoff, Sascha, 338*n*61
stem cells
of hydra, 116
longevity and, 125
meristematic cells and, 126
in sponges, 125
sterility, longevity and, 100–103
The Sting of the Wild, 307*n*3
stingrays, 246
stonefish, 246
stonustoxin, 246, 247
Strauss, Mark, 310*n*13
strength, mutations producing, 18
stress
in animals after September 11 terrorist attacks, 43
on deep sea organisms, 124–125
emotional intelligence and, 265
memories of trauma, 267–273
in *R. varieornatus,* 194
simplicity-stress-survivability equation, 125–129
studying whale EBC to determine, 45
studying whale poop to determine, 43–44
strongest. *see* toughest
Strudwick, Paul, 341–342*n*10
Stuckey, Alex, 310*n*19
Suncus etruscus. see Etruscan shrew
Super Hornet (jet), 133
superlatives
children's enchantment with, 2–3
in every area, 9
humans' fascination with, 3–4
multiple, in individual organisms, 6
recent scientific interest in, 8–9
science's lack of focus on, 8
in understanding human potential, 10–11
in understanding life and the universe, 9–10
superorganisms
ant colonies as, 215–217
defined, 215
energy use by, 216
SWARM (Search with Aerial RC Multirotor), 276
Swift, Jonathan, 19
Switek, Brian, 323*n*21
Synanceia (stonefishes), 246
synthetic spider silk, 250–251

T

Taiji dolphin slaughter, 263–264
tailings (radioactive waste), 58–59
Takacs, Zoltan, 338*n*60
tallest
investigating, 296
trees, 48–49, 296
Taman Negara National Park, 307*n*2
Tanner (bottlenose dolphin), 259–260
tarahiki, 40
tarantula hawk, 307*n*3
tardigrades (water bears)
collecting, 295
genes to fix DNA damage in, 194–195

toughness of, 190–194
Tarvin, Rebecca, 232
Taung Child, 322–323*n*24
taxonomic nomenclature, assumptions in, 143
Tears of the Cheetah (O'Brien), 138
Teeling, Emma, 197
Te Ika-a-Maui, 37
telomeres, 6, 113–114
Temnothorax albipennis (ant species), 216
Tero, Atsushi, 277
testicles, of howler monkey, 161
thale cress *(Arabidopsis thaliana),* 291
Thaumoctopus mimicus (mimic octopus), 273
"Thinking Like a Mountain" (Leopold), 108
Thompson, Grant, 337*n*41
Thompson, Ken, 317*n*41
threatened species, preserving, 28–29
three-toed sloth *(Bradypus),* 9, 204–206
Thunnus thynnus. see Atlantic bluefin tuna
tiger mosquito, 81, 317*n*38
tiger moth *(Bertholdia trigona),* 180, 330*n*55
tiger pistol shrimp, 170
tigers, 135
Tiktaalik, 196
Tinkerbella nana (wasp species), 78, 87
tirofiban, 244
tobacco
  cancers caused by, 228
  in fight against cancer, 228–231
  spread of, 335–336*n*24
tobacco/cigarette companies, 228–229, 335*nn*20–23
Tobin, Kate, 319*n*14
Todtling, Josef, 3
Tompkins, Peter, 288
Torres, Leigh, 37–40
Torres-Florez, Juan Pablo, 312*n*40
touch-me-not plant *(Mimosa pudica),* 290
toughest, 189–217
  how animal with longest know genome might unleash our X-powers, 200–203
  how slowest-evolving organism helps preserve biodiversity, 195–200
  how tardigrades can help reach other planets, 193–195
  investigating, 296
  what ants teach us about being super, 212–217
  why beetles could help feed the world, 207–212
  why slowest mammal is also one of toughest, 203–207
Tovey, Craig, 216–217
Towsey, Michael, 308*n*7
toxins database, 239
Trans-Tasman Resources, 39–40
trauma, memories of, 267–273
Travels into Several Remote Nations of the World (Swift), 19
tree of life, 62–65
trees. *see also specific trees*
  common genes of humans and, 115
  common genes of other organisms and, 115
  determining age of, 96–97, 110–112
  number of, 48
  oldest, 109–110
  population of, 313*n*50
  protection of, 106–108
  redwoods, 48–49
  sequoia, 52
  tallest, 48–49, 296
  tree-sitters, 47
triploids, 100–103
true hemlock *(Conium maculatum),* 226
Trypanosoma cruzi (parasite), 221
tsetse flies, 221
Tsuda, Dick, 316*n*32
Tucker, Ken, 345*n*51
Turnbull, Bert, 248
Tursiops truncatus (common bottlenose dolphin), 262–263, 265
Tushingham, Shannon, 335–336*n*24
two-toed sloths *(Choloepus),* 205, 206
type 1 diabetes, 246
Tyrannosaurus rex/T. rex, 172, 173

U

Ukrainian quagga mussel, 79
ultrasound, 166–167, 170, 178–181
UnDisciplined, 308*n*11
ungulates, 106–107
United Nations' Food and Agriculture Organization, 210
universe
- expansion of, 309*n*17
- size of, 10

University of Aarhus, 183
University of California at Irvine, 255
University of Hawaii at Manoa, 279
University of Kalyani, 226
University of Massachusetts Large Pelagics Research Center (Tuna Lab), 154–155
University of Modena, 192
University of Twente, 170
University of Virginia, 235
University of Washington, 48–49
unsupervised machine learning, 343*n*32
Upfield, Arthur, 324–325*n*41
Uppsala University, 235–236
US Department of Agriculture, 223
US Forest Service, 53, 118, 225
US Institute of Forest Genetics, 113
US Navy, 156–157, 184, 185, 328*n*31
Utah State University, 9

V

Vane, John, 241, 242
van Leeuwenhoek, Antonie, 59, 73, 314*n*1, 316*n*22
Van Pelt, Robert, 313–314*n*51
Van Strien, Jan, 238
Van Volkenburgh, Elizabeth, 288–289, 292
velociraptors, 189
Venezuelan red howler *(Alouatta seniculus),* 160–161
Venkatesh, Byrappa, 197–199
venom-based medicines, 239–244
Venomics Project, 239, 241
Venter, Craig, 70, 71
Veratrum californicum, 223–226
Vermeer, Johannes, 73, 316*n*22
vertebrate
- slowest-evolving, 196
- smallest, 82–87

vervet monkeys, 237
Virunga Mountain Gorilla, 138–139
vision, in ants, 283, 344*n*43
von Mutius, Erika, 60
von Tschermark, Erich, 180
Voss, Randal, 202
Voyager 1 space probe, 10

W

Wada, Naomi, 322*n*3
Wagner, Eric, 330*n*60
Wallace, David Foster, 210–211
Walsh, Bryan, 343*n*34
warrior wasp, 307*n*3
Warwick Elementary School, 3
Washington State Department of Natural Resources, 53
wasps, 78, 87, 245–246
water bears. *see* tardigrades
water flea *(Daphnia pulex),* 69
water hemlock *(Cicuta douglasii),* 224
water snail *(Ammonicera minortalis),* 93–94
web crawlers, 281–282
Webster, Kevin, 66
Weisman, Alan, 191–192
Western Pacific gray whales, 312*n*35
whales. *see also specific types of whales*
- after September 11 terrorist attacks, 43
- common genes of humans and, 115
- exhaled breath condensate of, 44–46
- impact of maritime traffic on health of, 43, 312*n*38
- migrations of, 312*n*40
- slaughter of, 186–187
- sounds made by, 165
- studying poop of, 42–44

Wheeler, William Morton, 215
White, Nicholas, 340*n*85
White, Thomas, 265
White, Will, 329*n*41
white-faced capuchin monkeys, 237–238

Whitehead, Hal, 330*n*58
White Mountains (California), 108–113
Why Size Matters (Bonner), 19–20, 168
wildlife reserves, in South Africa, 268–269
Williams, David, 240
Williams, Ken, 61–62, 68, 93–94
Wilson, Leah, 345*n*51
Windmill, James, 181, 330*n*52
Winfield, Alan, 318*n*56
Woese, Carl, 62–63
Wolbachia-infected mosquitoes, 257
Wolmarans, Riaan, 322–323*n*24
wolves, 105, 108
Wong, Ee Phin, 270
World Health Organization, 254
World Toxin Bank, 338*n*60
The World Without Us (Weisman), 191–192
Wright, Jonathan, 147, 148
Wright, Orville, 149, 324*n*37
Wright, Wilbur, 149, 324*n*37
Wystrach, Antoine, 344*n*46

X

Xenopus, 7
Xenopus laevis (African clawed frog), 6
Xie, Jie, 308*n*7

Y

Yellowstone National Park, 105
Yong, Ed, 310*n*17
Yoon, Carol Kaesuk, 327*n*18
Young, Truman, 311*n*28
Your Inner Fish (Shubin), 196
Yucatan black howler *(Alouatta pigra)*, 160
Yuhas, Daisy, 339*n*74

Z

Zhang, Jinglan, 308*n*7
ziconotide, 248
Zielinski, Alex, 340*n*92
Zielinski, Sarah, 320*n*18
Zigadenus gramineus (grassy deathcamas), 223
Zoological Society of London, 8–9, 16
zoos, 268, 269
Zuri (elephant), 1–2, 11, 21, 24–25, 272, 307*n*1

ABOUT THE AUTHOR

Photo by Keith Johnson

MATTHEW D. LAPLANTE is an author, journalist, advocate for educational equity, and associate professor of journalism at Utah State University. As a journalist, he has reported from more than a dozen nations on subjects including ritual infanticide in Northern Africa, war in the Middle East, gang violence in Central America, and the legacy of genocide in Southeast Asia. He is a past recipient of the Kavli Award for science reporting and the Ancil Payne Award for ethics in journalism, and was a finalist for the Deborah Howell Award for feature writing, the Hillman Prize for social justice reporting, and the Livingston Award for young journalists. The books he has co-written include *Inheritance*, with geneticist Sharon Moalem, and the Nautilus Award–winning *Longevity Plan*, with John D. Day and Jane Ann Day. *Superlative: The Biology of Extremes* is his first solo book. LaPlante lives in Salt Lake City, and skis in Big Cottonwood Canyon, with his wife, Heidi, and daughter, Spike.